Environmental Policymaking

Environmental Policymaking

Assessing the Use of Alternative Policy Instruments

Edited by
Michael T. Hatch

State University of New York Press

Published by
State University of New York Press, Albany

Printed in the United States of America

For information, address State University of New York Press,
194 Washington Avenue, Suite 305, Albany, NY 12210-2384

Production by Judy Block
Marketing by Michael Campochiaro

Library of Congress Cataloging-in-Publication Data

Environmental policymaking : assessing the use of alternative policy instruments / edited by Michael T. Hatch
p. cm.
Includes bibliographical references and index.
ISBN-13 978-0-7914-6347-5 (alk. paper) — 978-0-7914-6348-2 (pbk. : alk. paper)
ISBN 0-7914-6347-8 (alk. paper) — 0-7914-6348-6 (pbk. : alk. paper)
1. Environmental policy. I. Hatch, Michael T., 1945–

GE170.E57695 2005
363.7'0561—dc22 2004043480

10 9 8 7 6 5 4 3 2 1

To Carol, Benjamin, and Becca

and

In memory of Ernie: teacher, mentor, friend

Contents

Illustrations

Figures

Tables

CHAPTER ONE

Assessing Environmental Policy Instruments

An Introduction

MICHAEL T. HATCH

Environmental policymaking appears to be at an important juncture as we enter the twenty-first century. While environmental protection continues to hold a prominent position on the political agenda of most industrial democracies the methods employed in the pursuit of this objective are often highly contested. The traditional forms of environmental regulation initially adopted in response to concerns about environmental pollution generally took a so-called command-and-control regulatory approach, direct government regulations that require certain types of behavior, either by prescribing uniform environmental standards or the specific process or technology that must be used to be in compliance. In recent years, however, attempts to address environmental problems through such an approach have encountered ever greater resistance. In the view of many, these instruments have proven either inappropriate or ineffective when confronted with a fundamentally different world from that existing at the time of environmental awakening in the late 1960s. As a consequence, there has been a growing interest in the use of different types of policy instruments ranging from green taxes and tradable permits to eco-audits and eco-labeling. The chapters in this volume provide case studies that generally chronicle this turn to alternative policy instruments.

Factors influencing the search for alternative policy instruments

The growing interest in the use of different types of policy instruments has been driven by several separate but related phenomena. Common to most all is a general disaffection with traditional command and control instruments. The critique has taken various forms. One of the fundamental premises underlying a study by Chertow and Esty (1997) calling for a next generation of environmental policies in the United States is the ineffectiveness of traditional command and control regulations. While acknowledging that laws enacted in response to growing environmental activism in the late 1960s benefited the environment, they assert that further progress through the command and control regulatory approach is limited: many current environmental problems are different and some of the residual problems (automobile exhaust, agricultural runoff, loss of habitat to suburbanization) cannot be addressed effectively simply by further tightening of emissions standards (Esty & Chertow, 1997, p. 2). As argued by Elliot (1997, pp. 72–73), command and control approaches were developed to regulate large industrial polluters (such big dirties as power plants, refineries, chemical plants, and the automobile industry). The sources of environmental problems looming today are often smaller, more diffuse, and thus less amenable to traditional instruments. In other words, while certain sources of pollution may be overregulated, other sources are subject to little or no regulation.

Claims of overregulation are closely associated with concerns about the high cost or economic inefficiency of direct environmental regulation. By their very nature, command and control instruments are inflexible, imposing uniform emissions standards or technologies, irrespective of the varying conditions confronting individual firms as well as inefficiencies and costs these differences may generate. Moreover, by mandating specific technologies, the development of more effective or efficient technologies is discouraged; by prescribing specific emissions levels, there is little incentive to reduce below those levels, further serving to inhibit technological innovation (Golub, 1998b, pp. 2–4; Cohen, 1997, pp. 117–118). Important side effects of policies viewed as too costly or burdensome are the risk that the public support (so critical in the politics of environmental protection) could be undermined, investment in oversight and enforcement mechanisms restricted, and that noncompliance could become more problematic as administrative capabilities are diminished (Golub, 1998b, p. 4; Esty & Chertow, 1997, p. 6). Exacerbating these problems is the adversarial and legalistic nature of traditional regulatory strategies, which slow the formulation process and impede effective implementation and enforcement.

Concerns about economic competitiveness are often associated with claims of economic inefficiencies. The increasing globalization of economic activity in recent years has made more explicit the link between continued economic prosperity at home and the maintenance of international competitiveness. In a world of globalizing economic relations, where even local environmental problems now have important international economic ramifications, fears about the impact of environmental regulation on economic growth and competitiveness have heightened considerably.

Conventional wisdom holds that environmental regulation will have adverse effects on international competitiveness, economic growth and employment as production costs associated with environmental regulations reduce competitiveness in global markets. Rigorous environmental standards push industry abroad (industrial flight) and developing countries attempting to compete provide pollution havens. Confronted with such threats, there is the additional fear that policymakers will weaken environmental rules (the so-called race to the bottom). A number of studies attempting to test these hypotheses find little evidence to support such claims (OECD, 1993, 1997; Atkinson, 1996). Nonetheless, concerns about the loss of competitiveness due to government regulations continue to shape the views of many in the public and private sectors. Indicative of such concerns was the creation of the Competitiveness Council under the tutelage of Vice President Quayle during the first Bush administration (CPC, 1991), the debate in Germany about the country's ability to compete as a business site (Standortbericht, 1993) and the Molitar Report, a European Union study written by a group of independent experts expressing alarm about the potential threat to Europe's competitiveness posed by excessive environmental regulation (CEC, 1995; Golub, 1998a). In sum, though there is little evidence environmental regulation adversely impacts international competitiveness, such concerns continue to inform industry's approach to environmental policy. Regulatory standards are consistently opposed and voluntary approaches or market-based mechanisms such as emissions trading are, at least in the abstract, the preferred alternatives.

Taking this line of inquiry on the effect of environmental regulation on competitiveness one step further, some argue that the right type of regulatory instrument may, in fact, enhance a country's competitive advantage.

Variously characterized as the win-win or Porter thesis, and an important component of the ecological modernization paradigm that asserts a mutually supportive relationship between environmental protection and economic growth, (see, e.g., Blowers, 1998; Golub, 1998a; Hoerner, Miller, & Muller, 1995; Moore & Miller, 1994; Porter & van der Linde, 1995; Wallace, 1995), environmental regulations—properly constructed—can trigger innovation and encourage firms to upgrade their technology. The resulting operational changes and improved production processes often lead to greater productivity and higher product quality at less expense. Moreover, when a nation initiates environmental actions in advance of other countries, its companies can gain first mover or early mover benefits. That is, companies encouraged to develop less polluting technologies will enjoy competitive advantages in expanding markets abroad; other countries—driven by domestic demands for a cleaner environment and/or the process of trading up, whereby stricter environmental regulations are exported through such international institutions as the European Union or NAFTA (Vogel, 1995)—adopt similar standards. Properly constructed regulatory instruments that aim at outcomes instead of methods stimulate innovation rather than locking in the use of specific environmental technologies. In other words, market-based instruments or voluntary approaches, if sufficiently rigorous and adhered to, should provide the type of regulatory framework conducive to enhancing competitiveness—one that leaves discretion and the initiative for innovation in the hands of industry.

Finally, to the degree the sustainable development paradigm informs policymakers' thinking about environmental policy, instruments that represent alternatives to command and control regulation should begin to have a larger presence in the regulatory approaches of nations. According to Agenda 21 (UN, 1992), the document adopted at the Earth Summit in Brazil to provide a blueprint for action for global sustainable development into the twenty-first century, market-based mechanisms and such voluntary instruments as environmental labeling, self-regulation, and eco-auditing by industry should be given prominent roles in national strategies to encourage changes in nonsustainable consumption patterns. In terms of actual programs, the Dutch government, for example, initiated the National Environmental Policy Plan (NEPP) in 1989. Informed by the Brundtland Report (WCED, 1987)—and anticipating many elements in Agenda 21—the NEPP explicitly embraced the goal of sustainable development and, in so doing, shifted its regulatory approach from that of direct regulation to voluntary agreements negotiated between the state and private actors (for details, see Liefferink, 1999). The concept of sustainability was also at the center of the European Union's Fifth Environmental program (1992–2000) which emphasized, among other things, the need to broaden the range of environmental tools to include such instruments as environmental taxes and voluntary agreements (EU, 1998).

All told, disaffection with command and control regulation—whether based on concerns about economic efficiency, international competitiveness, environmental effectiveness, or the need for an approach that encourages sustainable development—has resulted in a search for alternative policy instruments believed to provide potential antidotes to such shortcomings.

Types of regulatory instruments and their assessment

For the purposes of this study, regulation is understood as "any attempt by the government to influence the behavior of citizens, corporations, or subgovernments" (Cohen, 1997, p. 110). There are a number of highly involved, perhaps overly complex, taxonomies or classifications suggested for regulatory instruments (see, e.g., Vedung, 1998). From our perspective, the most fundamental and salient distinction to highlight is that between mandatory and voluntary policy instruments. With the former, the regulated party's options are generally quite limited and the abrogation of mandated actions ultimately carries the possibility of legal sanction. Voluntary instruments, on the other hand, are most often nonbinding and allow considerable flexibility or discretion. Moreover, depending on the instrument involved, one might anticipate the development of a different type of political dynamic. For example, certain instruments may encourage a more adversarial approach that slows, if not hinders, the adoption and implementation of policy, whereas other instruments may be more conducive to a collaborative policy process that facilitates prompt implementation.

Included in the category of voluntary policy instruments are environmental or eco-labels, eco-audits and voluntary agreements. Among the distinguishing features attributed to each are the following:

- Environmental labeling programs formulate a set of production or performance criteria products must meet if they desire to carry the eco-label. Products certified as meeting these criteria embody more environmentally benign production and consumption practices than those of their noncertified competitors and offer consumers a choice based on ecological considerations. Eco-labels, in other words, create incentives for the innovation of more environmentally sound products or production practices by providing information upon which the environmentally conscious consumer can then act. Given the voluntary nature of this instrument, proponents argue that eco-labels are relatively light in terms of the amount of public expenditure, management, and oversight are concerned, and as such, are rather easy to introduce and implement.

- Similarly, eco-audits are voluntary arrangements that provide consumers information about environmental management practices. For firms that choose to adopt specified standards for environmental management, certified participation in such programs is designed to foster better relations with customers, suppliers, and stakeholders as well as employees. Moreover, through this process of self-evaluation, a firm may discover ways of doing things more efficiently, thereby reducing its ecological footprint. From the perspective of the public sector, such arrangements promise environmental benefits without the high administrative costs that accompany direct regulation.
- Voluntary agreements take a variety of forms ranging from industry covenants that are legally binding to informal declarations of intent. At its core, however, a voluntary agreement is "an agreement between government and industry to facilitate voluntary action with a desirable social outcome, which is encouraged by government, to be undertaken by the participant based on the participant's self interest" (Storey, 1997, p. 11). In contrast to command and control regulation, firms may voluntarily agree to certain emissions targets, but they have much greater flexibility in terms of method and timing, thereby encouraging greater efficiency and innovation (Golub, 1998b, p. 5). In other words, voluntary agreements help achieve environmental objectives at lower costs and often more quickly, given the more collaborative nature of the policy process that such an approach implies.

Within the category of mandatory policy instruments, the most frequently cited are market-based taxes and tradable permits, environmental impact assessments and, of course, command and control regulation:

- Green taxes are charges assessed on an amount of pollution that a firm or product generates. Faced with the direct costs of their polluting activities, firms have an incentive to control pollution. At the same time, they are free to choose the most efficient reduction methods. In addition, green taxes provide ongoing incentives to find the most efficient reduction technologies in order to lower or avoid the tax. Finding the proper level of taxation, however, is critical to the effectiveness of the instrument because it is difficult to anticipate exactly how much pollution reduction will result from any given tax (see Stavins & Whitehead, 1997, pp. 106–107).

- Emissions trading too employs the price mechanism to internalize the costs of pollution, thus encouraging both the static and dynamic efficiencies that lead to ongoing pollution reduction. In contrast to green taxes, however, tradable permits avoid having to predict the appropriate level of taxation required to reach the reduction goals. Under an emissions trading system, policymakers establish an overall target of emissions allowed for an industry, area, or country. Permits representing shares of the total emissions allowed are then allocated to each company. Firms that reduce their emissions below the allotted levels can sell the surplus to firms whose emissions exceed their permits or bank them for the future. Companies that exceed their permitted limits must purchase permits from other firms or face legal sanctions (Stavins & Whitehead, 1997, p. 107; Golub, 1998b, p. 5).

- Environmental impact statements (EISs) represent a process designed to identify, evaluate, predict, and mitigate the effects of proposed developments before decisions are taken or projects initiated. Generally mandated by law, they are to provide information on the environmental effects, risks, and consequences of development proposals. By explicitly integrating science into the decisionmaking process, EISs are said to facilitate the inclusion of environmentally sound and sustainable options in proposed projects (see, e.g., Sadler, 1996).

- There are generally two types of direct or command-and-control regulatory instruments, technology-based and performance-based. Technology-based regulations typically prescribe the use of specific equipment, processes or procedures, whereas performance-based standards specify the level of pollutant emissions allowed (Stavins, 1997, p. 8). Proponents argue both approaches are effective in achieving their specified environmental objectives. An added benefit of technology-based regulations is that compliance and oversight are made much easier; moreover, practices incorporating such principles as Best Available Technology are said to provide greater flexibility than commonly believed (Dente, 1995, p. 18). Similar arguments about the virtues of flexibility made by proponents of voluntary and market-based instruments are made in the case of performance-based regulations as well, since the regulated entities have considerable discretion in that only the levels of emissions are prescribed, not the methods allowed to achieve them.

All told, contending assertions are made about the relative virtues of these policy instruments. Similar to the critiques of command and control cited earlier, however, the claims made by advocates of voluntary and market-based instruments have not gone unchallenged. Supporters of command and control express concerns that the increased flexibility allowed by voluntary approaches will lower overall environmental protection. In the words of the Sierra Club's executive director, for example, given the opportunity to cheat on environmental standards, some people do and others will follow, "Consistent, strictly enforced standards are the locks that protect our environment. Voluntary measures and self-policing sound appealing, but they aren't enough to keep law and order on the environment frontier" (Pope, 1999, p. 17).

Proponents of economic instruments are skeptical voluntary arrangements can produce economically efficient results since they are based on political bargains rather than price signals. Similarly, green taxes, when implemented in the real world, reflect the influence of powerful political forces as much as market forces and limit the environmental effectiveness of the instrument. At a more fundamental level, it is asserted that such market-based instruments as emissions trading essentially represent a right to pollute. In other words, the conflicting claims about these various policy instruments are very much in dispute. As such, it would be prudent to treat these assertions as quasi hypotheses in need of testing. This will be the central focus of the chapters that follow.

The hypothesized claims about the merits (and shortcomings) of these instruments are based, at least in part, on differing views regarding the priorities that ought to govern environmental policy. Such differences, in turn, are often reflected in the criteria employed when arguing the relative usefulness of particular instruments. Among the criteria that frequently inform such analyses are the following:

- Environmental effectiveness is perhaps the most common criterion used to evaluate policy instruments. Its precise meaning, however, is not self-evident. To some, it means eliminating the problem to be addressed. Others favor a definition that emphasizes some physical improvement in the environment. But given the highly complex nature of many environmental problems, as well as the lag times that frequently occur between action and impact, it is often difficult to establish causal links between any physical changes that may occur and the policy instrument. In the absence of reliable data, surrogates such as measuring the degree to which targets established by a policy instrument are met and/or the use of counterfactuals have been proposed (see Keohane, Haas & Levy, 1993; Helm & Sprinz, 2000).
- Economic efficiency emphasizes the cost effectiveness with which an instrument is able to achieve its policy objective.

- The choice of policy instruments is rarely the result of a purely technical selection process. It is the outcome of a political process that engages a myriad of actors with competing interests and priorities. As a consequence, the performance of that instrument may not achieve the idealized outcome hypothesized for other instruments in the abstract, ignoring the fact that the policies required to achieve that outcome are not politically viable. The criterion of political efficiency highlights the value of feasibility and second-best solutions when evaluating policy instruments (see Müller, 1999). It too suggests attention ought to be paid to the possibility that certain policy instruments may help realign interests and generate much needed political support for policy initiatives.

- As the growing literature on implementation and compliance in environmental policy suggests (see, e.g., Mitchell, 1994; Brown Weiss & Jacobson, 1998; Victor, Raustiala, & Skolnikoff, 1998), there is no guarantee that policies, once adopted, will be implemented as intended. The criterion of administrative efficacy points to the possibility that the design of certain types of instruments may enhance the implementation of and compliance with adopted policies. For example, regulations that mandate specific technologies may facilitate compliance and oversight, since it is fairly easy to detect whether or not the technology is employed. More collaborative approaches may also engender compliance as participants in the process feel more invested in the outcome.

- An important consideration for some when evaluating policy instruments is their effect on technological innovation—that is, the ability of various instruments to encourage rather than impede the development of technologies that reduce or prevent pollution (see Esty & Chertow, 1997, p. 12; Norberg-Bohm, 1999).

To conclude, a number of claims made about the virtues of the various policy instruments discussed above are most often grounded in their greater economic efficiency, environmental effectiveness, or technology-forcing capabilities. In the analyses that follow, efforts will be made to assess the usefulness of each policy instrument using the criteria just discussed. At the same time, since these claims are very much in dispute, they will be treated as quasi hypotheses to be tested.

The method employed for this testing are case studies that provide detailed analyses of the individual policy instruments in their practical application. Single cases, of course, neither confirm nor disprove the general validity of hypotheses. What they can do, however, is help refine the understanding of specific conditions under which an instrument may be more or less useful (i.e., move beyond hypothesized correlations to observable empirical relationships), thereby facilitating a more thoughtful elaboration of hypotheses. At the same time, these case studies are intended to contribute to the cumulative process of constructing a body of literature that will assist in the eventual confirmation of broader generalizations.

Overview of the book

The most fundamental differentiation in the type of regulatory instruments employed in environmental protection is between voluntary and mandatory instruments. The chapters in this volume are organized accordingly: chapters 2–5 provide case studies in the use of voluntary instruments, beginning with one of the earliest experiments with this type of approach. Chapters 6–10 focus on the application of several mandatory instruments in various national and international settings.

During the 1990s, environmental labels became increasingly popular throughout the world. However, as detailed by Edda Müller in chapter 2, one of the earliest efforts to employ this instrument began two decades earlier with the adoption of the Blue Angel program in Germany. Drawing on lessons from this eco-labeling program, Müller addresses three questions: First, is the environmental label really a light, less complex tool that is easily implemented? Second, how effectively are its economic and ecological objectives achieved? And third, what is the role of eco-labeling in an environmental policy mix that aims at stimulating innovation and diffusion of technical and social change?

In chapter 3, Ronnie D. Lipschutz continues the discussion on voluntary regulatory practices, but from a perspective that focuses on the shift of regulatory power to private, nongovernmental actors. Lipschutz argues that for reasons linked to globalization, there has been a gradual decline of public sector involvement in addressing various social problems, and growing difficulties in trying to devise cooperative public international conventions. As a result, nonprofit actors are engaged in a growing number of semipublic and private regulatory projects. To a large degree, such projects are the work of a global civil society and serve as the basis for a global system of democratized regulation for the rest of us. Lipschutz considers the general problematic of rule and rules in world politics, the sources of demand for global regulatory arrangements, and the activities of global civil society organizations and movements to meet these demands. With focus on the forestry sector, he provides a general discussion of the privatization of regulatory authority and involvement by civil society actors and describes several such initiatives currently underway.

In an effort to avoid regulation by the state, industries are suggesting alternative means of promoting environmental management within the corporate sector. These include such voluntary environmental management systems as the ISO 14001. In recent years, Japanese corporations have become world leaders in the number of companies that have applied for and received this certification. Multinational companies are now starting to demand similar certification of suppliers, many of which are based elsewhere in Asia. In chapter 4, Miranda Schreurs and Eric Welsh examine why Japanese corporations have suddenly become so interested in ISO 14001 and what it might mean for environmental protection in the Asian region.

In addition to eco-labeling and eco-auditing arrangements, governments and industry increasingly have entered into dialogues resulting in voluntary agreements. Proponents argue that voluntary agreements provide firms greater flexibility in the method and timing of activities, thus reducing the overall costs of environmental protection as well as stimulating innovation. Moreover, voluntary approaches are said to further understanding and trust between government and industry and promote faster implementation and compliance. Questions, however, have been raised about environmental effectiveness. Critics argue that voluntary agreements are not likely to move beyond what industry would have done in the absence of such commitments and in foregoing formal laws or ordinances, such commitments fall well short of what could have been achieved. In chapter 5, Michael T. Hatch sorts out the conflicting claims about the environmental effectiveness of voluntary agreements through an examination of what has become the centerpiece of Germany's strategy to reduce its CO_2 emissions 25 percent by the year 2005—a set of voluntary agreements concluded between the German government and industry in 1995 and subsequently revised in 1996, 2000, and 2001. Special attention is paid to the agreement between government and the electric utility industry—the sector responsible for approximately one-third of Germany's total CO_2 emissions.

Another part of Germany's effort to combat global warming was the adoption of a green tax. In 1999 Germany entered into an ecological tax reform. It was designed to increase energy taxes and reduce social security contributions in five steps up to 2003. It aimed at two goals simultaneously: to reduce energy consumption along with the accompanying emissions of climate gases while, at the same time, cut labor costs to decrease unemployment. This reform is highly controversial. It has been hailed as a central project of the modern age by some proponents and denounced by opponents as a misguided attempt to satisfy the government's need for increased revenue and an impediment to economic growth and employment. In chapter 6, Michael Kohlhaas and Bettina Meyer describe the cornerstones of the new law and address the most controversial issues from an economic and political point of view. They argue that not all deviation from the optimal design of an ecological tax reform means bad politics. The law had to balance competing objectives and interests and was subject to legal, technical, and political restrictions that could not be overcome in the short term.

Whereas the Hatch and Kohlhaas/Meyer chapters look at policy instruments employed nationally in response to the challenges posed by climate change, several instruments have been proposed at the international level that would assist in reducing GHG emissions: the so-called flexible mechanisms of Joint Implementation (JI), the Clean Development Mechanism (CDM), and emissions trading established in the Kyoto Protocol. In chapter 7, Andreas Oberheitmann focuses on two of these programs, JI and the CDM, and asks why many Annex I Parties and their private companies appear reluctant to participate in them. He asserts that among the reasons for this reluctance are uncertainties about the implications of JI and CDM for instruments employed in national climate policy as well as the potential obstacles to their effective operation. Oberheitmann then analyzes how these instruments relate to those employed in national climate policy—specifically, CO_2-taxes and voluntary agreements—and he looks at the questions raised about their compatibility with international agreements and provisions on subsidies governed by the WTO.

The flexible mechanisms proposed in the Kyoto Protocol essentially represent different types of trading arrangements in greenhouse gas emissions. In chapter 8, Gary C. Bryner examines the use of emissions trading in the United States, one of the few countries that has extensive experience with this instrument. Based on an analysis of the federal emissions trading established in the Clean Air Act and Southern California's South Coast Air Quality Management District's RECLAIM program, the strengths and limitations of emissions trading are discussed along with the implications of the U.S. experience for a greenhouse gas or carbon trading program.

Another area where the United States has been an innovator is in the use of environmental impact statements. Originally adopted with the passage of the National Environmental Policy Act (NEPA) in 1970, environmental impact assessments today are required in over one hundred countries—and many subnational bodies in federal systems—as well as in many international lending institutions and bilateral aid agencies. Yet, despite its widespread popularity, relatively little attention has been devoted to the use of this instrument in recent years. Walter A. Rosenbaum seeks to redress this situation in chapter 9 as he analyzes the evolution of Environmental Impact Statements (EISs) in the United States. NEPA's mandate to federal agencies to create environmental impact statements was intended to create a profound change in federal environmental policymaking, both procedurally and substantively. At least five impacts were predicted by proponents of EISs: (1) greater bureaucratic integration of science and environmentally-related policy; (2) more environmentally benign federal policies; (3) an early warning system for environmentalists about the implications of evolving agency policies; (4) development of an environmentally sensitive infrastructure within major environmental bureaucracies; and (5) alterations in agency cultures. Rosenbaum argues the most important impacts are those involving the integration of science and environmental values within agency structures and that the course of this integration has been profoundly influenced, in ways largely unanticipated, by the federal judiciary.

In chapter 10, Daniel H. Cole and Peter Z. Grossman revisit the debate about the relative efficiency of economic as opposed to command and control regulatory instruments. They question the assertion that economic forms of regulation such as effluent taxes and emissions trading are inevitably more efficient than traditional command-and-control regimes for environmental protection. They take issue with the general portrayal of command-and-control environmental regulations in the economic literature that focus almost exclusively on compliance costs, ignoring technological and institutional constraints that can significantly affect the comparative efficiency of alternative environmental policies. Their analysis of air pollution control under the U.S. Clean Air Act suggests that where abatement costs are relatively low and monitoring costs are relatively high, command-and-control is likely to be at least as efficient (and effective) as effluent taxes or a tradable emissions program.

References

Agenda 21: A blueprint for action for global sustainable development into the 21st century. (1992). New York: Author.

Atkinson, R. D. (1996). International differences in environmental compliance costs and United States manufacturing competitiveness. *International Environmental Affairs, 8*(2), 107–134.

Blowers, A. (1998). Power, participation and partnership: The limits of co-operative environmental management. In P. Glasbergen (Ed.), *Co-operative environmental governance: Public-private agreements as a policy strategy.* Dordrecht: Kluwer Academic Publishers.

Brown Weiss, E., & Jacobson, H. K. (Eds.). (1998). *Engaging countries: Strengthening compliance with international environmental accords.* Cambridge, MA: MIT Press.

Chertow, M. R., & Esty, D. C. (Eds.). (1997). *Thinking ecologically: The next generation of environmental policy.* New Haven, CT: Yale University Press.

Cohen, S. (1997). Employing Strategic Planning in Environmental Regulation. In S. Kamieniecki, G. A. Gonzalez & R. O. Vos (Eds.), *Flashpoints in environmental policymaking.* Albany, NY: State University of New York Press.

Commission of the European Communities (CEC). (1995). *Report of the Group of Independent Experts on Legislative and Administrative.* COM (95) 288, 21 June, Brussels.

Competitiveness Policy Council (CPC). (1991, March). *Building a competitive America.* First Annual Report to the President and Congress.

Dente, B. (1995). Introduction: The globalization of environmental policy and the search for new instruments. In B. Dente (Ed.), *Environmental policy in search of new instruments.* Dordrecht: Kluwer Academic Publishers.

Elliot, E. D. (1997). Toward ecological law and policy. In Chertow & Esty (Eds.).

Esty, D. C., & Chertow, M. R. (1997). Thinking ecologically: An introduction. In M. R. Chertow & D. C. Esty (Eds.).

European Union (EU). (1998). Decision No. 2179/98/EC of the European Parliament and of the Council of 24 September 1998 on the review of the European Community programme of policy and action in relation to the environment and sustainable development 'Towards sustainability.'

Golub, J. (1998a). Global competition and EU environmental policy: Introduction and overview. In J. Golub (Ed.), *Global competition and EU environmental policy* (pp. 1–33). London and New York: Routledge.

Golub, J. (1998b). New instruments for environmental policy in the EU. In J. Golub (Ed.), *New instruments for environmental policy in the EU* (pp. 1–29). London and New York: Routledge.

Helm, C., & Sprinz, D. (2000). Measuring the effectiveness of international environmental regimes. *Journal of Conflict Management, 44*(5), 630–652.

Hoerner, J. A., Miller, A. S.. & Muller, F. (1995, April). *Promoting growth and job creation through merging environmental technologies,* Research Report No. 95-03, Washington, DC: National Commission for Employment Policy.

Keohane, R. O., Haas, P. M., & Levy, M. A. (1993). The effectiveness of international environmental institutions. In P. M. Haas, R. O. Keohane, & M.A. Levy (Eds.), *Institutions for the earth: Sources of effective international environmental protection.* Cambridge, MA: MIT Press.

Liefferink, D. (1999). The Dutch national plan for sustainable society. In N. J. Vig & R. S. Axelrod (Eds.), *The global environment: Institutions, law, and policy* (pp. 256–278). Washington, DC: CQ Press.

Mitchell, R. B. (1994). *International pollution at sea: Environmental policy and treaty compliance.* Cambridge, MA: MIT Press.

Moore, C., & Miller, A. (1994). *Green gold: Japan, Germany, the United States and the race for environmental technology.* Boston: Beacon Press.

Müller, E. (1999). Ökonomische Effizienz und politische Effizienz in der Umweltpolitik. In E. Gawel & G. Lübbe-Wolff (Eds.), *Rationale Umweltpolitik-Rationales Umweltrecht* (pp. 00–00). Baden-Baden: Nomos Verlagsgesellschaft.

Norberg-Bohm, V. (1999). Stimulating "green" technological innovation: An analysis of alternative policy mechanisms. *Policy Sciences, 32,* 13–38.

Organization for Economic Cooperation and Development (OECD). (1993). *Summary report of the workshop on environmental policies and industrial competitiveness, 28–29 January 1993.* Paris: Author.

Organization for Economic Cooperation and Development (OECD). (1997). *Economic globalisation and the environment.* Paris: Author.

Pope, C. (1999, July–August). Corporate crime: The consequences of letting polluters police themselves. *Sierra,* 16–17.

Porter, M. E., & van der Linde, C. (1995). Toward a new conception of the environment-competitiveness relationship. *Journal of Economic Perspectives, 9*(4), 97–118.

Sadler, B. (1996, June). *Environmental assessment in a changing world: Evaluating practices to improve performance.* Final Report, International Study of the Effectiveness of Environmental Assessment.

Standortbericht. (1993, September). Report on securing Germany's future as a business site.

Stavins, R. N. (1997, January). Policy instruments for climate change: How can national governments address a global problem? Resources for the Future Discussion Paper 97–11.

Stavins, R., & Whitehead, B. (1997). Market-based environmental policies. In M. R. Chertow & D. C. Esty (Eds.).

Storey, M. (1997). *Voluntary agreements with industry.* Annex I Expert Group in the UNFCCC, Working Paper No. 8. Paris: OECD.

United Nations. (1992).

U.S. Congress, Senate. (1969). *National Environmental Policy Act of 1969.* 91st Congress, 1st session.

Vedung, E. (1998). Policy instruments: Topologies and theories. In M-L. Bememlmans-Vidic, R. C. Rist, & E. Vedung. *Carrots, sticks sermon: Policy instruments and their evaluation.* New Brunswick, NJ: Transaction Publishers.

Victor, D. G., Raustiala, K., & Skolnikoff, E. B. (Eds.). (1998). *The Implementation and Effectiveness of International Environmental Commitments.* Cambridge, MA: MIT Press.

Vogel, D. (1995). *Trading up: Consumer and environmental regulation in a global economy.* Cambridge, MA: Harvard University Press.

Wallace, D. (1995). *Environmental policy and industrial innovation: Strategies in Europe, the USA and Japan.* London: Earthscan.

World Commission on Environment and Development (WCED). (1997). *Our common future.* Oxford: Oxford University Press.

CHAPTER TWO

Environmental Labeling, Innovation, and the Toolbox of Environmental Policy

Lessons Learned from the German Blue Angel Program

EDDA MÜLLER

Introduction

Environmental labeling programs have become increasingly popular around the globe. They seem to fit perfectly into a sustainability strategy, which aims at reconciling economic, social, and ecological objectives by enabling and fostering innovation for more sustainable, resource efficient, and ecologically benign production and consumption patterns. In doing so, it gives high priority to cooperative, consensual, and voluntary actions, and self-determined individual environmentally conscious behavior.

The use of environmental labeling programs was recommended by the 1992 Earth Summit in Rio. Agenda 21, the United Nations program of action for sustainable development, explicitly provides that governments should encourage the expansion of environmental labeling in order to change consumption patterns (UN, 1992, chap. 4). Moreover, environmental labeling programs are mentioned as an important tool in several policy papers by the Commission on Sustainable Development (CSD) and Organization for Economic Cooperation and Development (OECD) to promote sustainable consumption patterns (OECD, 1991, 1997). They are accepted worldwide as an environmental management tool within the ISO 14000 family by ISO Standard 14021, which specifically addresses guidelines and procedures for a transparent and coherent environmental labeling scheme. The conclusions drawn at the 1998 Berlin Conference, Ecolabeling for a sustainable future, organized by OECD on the occasion of the twentieth anniversary of the German Blue Angel scheme, underlined the role of eco-labeling as a policy tool to promote environmentally preferable products and services (OECD, 1999). With the increase of eco-labeling programs worldwide, however, this tool also provokes criticism. Mainly developing countries perceive them as possible disguised trade barriers. The impact is noticeable. Governments attending the World Summit on Sustainable Development in Johannesburg in late summer 2002 could not agree to even mention eco-labeling among the consumer information tools when negotiating the chapter "Changing unsustainable patterns of consumption and production" of the Johannesburg Plan of Implementation (advanced unedited text, September 4, 2002).

This chapter discusses the role of environmental labeling in the overall toolbox of environmental policy aiming at stimulating innovation and technical change. It focuses on third party organized environmental labeling such as Type I labeling, ISO 14024 (Type II eco-labels are self-declared environmental claims by manufacturers and Type III lifecycle assessment-based eco-profiles). More specifically, it addresses the strategy, institutional framework conditions, capacities, and resources needed to successfully implement this tool.

The findings from recent environmental policy research support the thesis that innovation can best be promoted by a policy mix approach that combines several different environmental policy tools (see Jänicke & Weidner, 1995; Müller, 1997, 1999b; Kemp, 1997; Blazejczak, Edler, Hemmelskamp, & Jänicke, 1999). This calls for a refined design of political strategies and a careful evaluation of the potential and limits of environmental labeling as a tool to promote innovation and diffusion of technical and social change.

Traditionally, national governments used mainly top-down command-and-control instruments to improve the environmental quality of products. These tools, however, were largely unsuccessful in getting product-related pollution under control. One of the reasons was that state policy could not cope with the speed and complexity of the development and marketing of new, potentially harmful products by manufacturers. A second, and more important, reason was that national environmental policy got increasingly difficult to unilaterally impose product standards on national markets due to economic globalization and free trade requirements. By fostering innovation, environmental labeling promises not to harm economic growth and social welfare while, at the same time, provides solutions to protect the environment and save finite natural resources—all agreed priorities of the environmental policy agenda. But the popularity of voluntary labeling programs among politicians may also have something to do with the impression that the political decision to introduce such a program would be uncontroversial and rather easy compared to the introduction of other instruments such as taxes or binding standards. Moreover, the attractiveness of this kind of environmental policy tool may also be influenced by the perception that it is a flexible, self-running, and self-financed tool requiring few public resources for its implementation. Environmental policy researchers also praise the relatively low degree of complexity of eco-labeling (Kern, Jörgens, & Jänicke, 1999, p. 13).

The questions I would like to discuss in this chapter are threefold:

- Is the environmental label really a light, less complex tool that is easily implemented?
- How effectively are its economic and ecological objectives achieved?
- What is the role of environmental labeling in an environmental policy mix that aims at stimulating innovation and diffusion of technical and social change?

To answer these questions I will draw lessons from the German Blue Angel Program.

In making the case, the following three sections introduce the Blue Angel Program, look at the implementation problems, and discuss the effectiveness and impact of the Blue Angel Program on innovation and diffusion.

The Blue Angel Program: The idea, facts, and figures

"Ideas do not float freely" (Risse-Kappen, 1994)

The idea of using consumers' environmental awareness and competition among manufacturers to promote the environmental quality of consumer goods and products by the introduction of an environmental labeling program was announced in the Environmental Program of the German Federal Government in 1971 (Umweltprogramm der Bundesregierung, 1972, p. 79). The basic idea was simple. The environmental quality of consumer goods and other products should be improved by providing an economic incentive to manufacturers to develop products. In addition, through all stages of the product life cycle—choice of raw material, production, use and disposal—these products should be less harmful to the environment than conventional ones serving the same purpose. This so-called life-cycle-approach later guided eco-labeling activities and has been reproduced in the ISO 14021 standard. A label indicating the specific environmental improvement should be placed on the product to inform consumers. Manufacturers should be allowed to advertise their products with environmental arguments.

The general climate in the early 1970s provided evidence the idea could work. Opinion polls indicated a high degree of environmental awareness in the German population (Rat von Sachverstädigen für Umweltfragen, 1978, 446ff). The environmental program was also received positively by industry and unions (Müller, 1986, 89ff). The commonly agreed paradigm of this period was qualitative economic growth. By the time the idea finally materialized in 1977 with the political decision to start the program, the situation had changed fundamentally. The first oil price shock had pushed environmental policy into a defensive position. Most of the legislative command-and-control activities were frozen. The eco-labeling idea, therefore, was politically welcomed as an icebreaker and as an attempt to call upon the responsibility of consumers and manufacturers to reduce product-related pollution on a voluntary basis.

The German environmental label is made up of the sign of United Nations Environmental Program (UNEP). Its name derives from the blue figure of the United Nations logo with its arms spread, surrounded by a blue circle with laurel wreath. The United Nations (UN) granted permission to use the UN environmental logo for the purpose of the environmental labeling program under the condition that the German authorities would ensure proper rules and procedures to avoid any misuse or deviation from the objectives of the program. The Blue Angel combines the UN environmental logo with verbal explanations of the reasons of awarding. During the first ten years, the words environment friendly were used on the top line; it was subsequently replaced by the neutral term Environmental Label. On the occasion of the twenty-fifth anniversary of the program in 2003, it was changed to Blue Angel.

For a successful functioning of the labeling program, three goals had to be reached:

1. To be informative for, and easily understood by, consumers.
2. To reduce product-related environmental pollution through innovation and diffusion of environmentally improved products.
3. To provide sufficient economic incentive to manufacturers to respond to the program.

With the decision to use the familiar environmental logo of UNEP rather than invent a new logo, the initiators of the Blue Angel program hoped for a good start in reaching the first goal. As will be shown later, the information task turned out to be much more complex and difficult.

The second and third goals were very much interrelated. Environmental targets could not be defined independently from the already existing potential and best available technology to start the process. As a consequence, innovation should be reached in stages by terminating Blue Angel criteria after three years (later extended to four years), the intention being to then revise and strengthen them in line with the rate of diffusion of best products on the market. The practical work, therefore, had to concentrate on the economic incentive for manufacturers. "To assist consumers to make informed choices" (Agenda 21) was in this sense the intermediate rather than final goal of the exercise. Mainly the retailer as a third partner had to play a role in the process. Retailers had to be willing to enlist labeled products and to reserve valuable space for it on their shopping shelves. We will see later that these conditions not only limited the scope of the labeling program in promoting innovative environmental friendly products; it also led to misunderstandings and criticism mainly from environmentally engaged groups and consumer organizations. In sum, the idea to use the environmental awareness of consumers and competition among manufacturers to promote environmental goals was far from free floating, nor was it devoid of criticism from surprising sources. Before turning to the implementation problems that arose, however, we need to understand the institutional framework, procedures and work process of the Blue Angel program.

Blue Angel is a voluntary instrument managed in public-private partnership. The institutional setting divides the roles as follows (RAL, 2000):

- The maintenance of the ambition of the program—by defining targets, thresholds and levels of standards, the prevention of abuses, the political representation and the verification of difficult or production-site related questions of applications—is the function of public authorities. The Federal Environmental Agency (UBA) formulates drafts of the Blue Angel awarding criteria, reviews new proposals and prepares proposals for the revision of existing Blue Angel criteria. It assists the private certifying institute in the verification of applications and acts on behalf of or together with the Federal Ministry for the Environment (BMU) to protect and represent the program. Authorities of the federal states are involved in checking applications related to environmental problems deriving from the production facilities of manufacturers.

- The elaboration and negotiation of the details of the awarding criteria is a joint venture between public experts, private experts and involved private stakeholders. Hearings are held with participants from industry, consumer and environmental associations, and, if necessary, scientists, representatives of testing institutes or other specialists. Here, the draft proposals of the Federal Environmental Agency are discussed prior to the final decision.
- The decision-making power is entirely in the hands of an independent body, the Environmental Label Jury. It has fourteen members with voting rights. Twelve experts represent science, industry and trade, unions, consumers, environmental associations, the churches, municipalities and the media. Also having membership on the Jury are the federal states, represented by the present and past year's chair of the Permanent Conference of the Ministers for the Environment. Representatives from the BMU, UBA and the certifying institute attend the Jury meetings with non-voting rights. The Jury decides by majority vote and is not obliged to follow the results of the hearings. The plenary of the Jury convenes twice a year.
- The entire program is administratively and financially managed by a private institution. RAL—German Institute for Quality Assurance and Certification—manages the process, organizes the hearings, examines the applications and concludes the contracts with manufacturers. It acts as the certifying institute.

Concerning the procedures, two phases have to be distinguished:

Phase 1: Selection of product categories, elaboration, and definition of awarding criteria for product categories, the so-called Basic Criteria;

Phase 2: Examination of applications for the conclusion of contracts on the use of the Blue Angel for specific products by manufacturers.

Both phases are very demanding in terms of technical expertise, resources, and time needed to do the job. The first phase, however, is perhaps most interesting for the judgments made about the possible impact of environmental labeling programs on technological innovation. In this regard, the nondecisions—those product categories for which awarding criteria have not been defined after a comprehensive analysis—merit our attention as well.

Decisions in Phase 1 have to be prepared on the selection of an appropriate product category as well as on the concrete definition of the product category. With the submission of a new proposal to the Federal Environmental Agency, an explanation has to be provided as to why and how the environmental quality of the proposed product category shows advantages compared to other products serving the same purpose. Possible advantages depend to a large extent on the understanding of what same purpose means. Does a sailboat serve the same purpose as a motorboat? Does a mechanical pipe cleaning device serve the same purpose as chemical pipe cleaners? In the first example, the Environmental Jury said no, but it voted yes in the case of the second. Again, to become effective, the label must be economically rewarding for manufacturers. Therefore products must not only theoretically but also practically compete with other products. In the case of the sailboat, the Jury assumed that a Blue Angel would not provide new information to a customer wanting to buy a boat. The purchaser of a motorboat would have made up his mind beforehand and would not change his decision because the Blue Angel had been awarded to sailboats. Therefore, it would not help the manufacturer increase his market share to the detriment of motorboat producers.

These differentiated judgements, which precede decisions on the awarding of a Blue Angel, are not easily communicated. The seemingly simple information tool turns out to suffer under the iceberg syndrome. Only the tip can be seen, and the larger part of the preparatory work remains under the surface.

The same is true for the preparation of the Basic Criteria, including requirements and verification criteria for selected product categories. The examination, done by the experts of the Federal Environmental Agency, includes several steps. A comprehensive analysis of the environmental aspects of all phases of the product life cycle is carried out and the ecological damages caused by the products in single phases in various areas of the environment are examined. Problems are evaluated in light of the technical feasibility of avoiding or reducing them. Here again, the reference is not the theoretically feasible solution. Rather, the criteria are defined in such a way that at least one domestic or foreign manufacturer will be able to bring the improved product to the market. The characteristics of possible substitutes for harmful substances must be known to avoid replacing one evil by another. Most important, the requirements and verification procedures have to be defined in such a way that tests are replicable and results quantifiable, or can be evaluated in some other objective, transparent manner. In this sense, preparation for awarding criteria is much more demanding than the comparative testing of goods, when no definitive targets are needed but differences can be evaluated by comparison. Although the awarding criteria emphasize the environmental quality of a product, safety and serviceability aspects cannot be ignored. Relevant technical standards or codes, therefore, will be included in the awarding criteria.

As important as these more technical problems is the objective of avoiding any inflation of labels for a given product category. That is why work on the Blue Angel concentrated on environmentally harmful products, whereas harmless product categories were left aside. It also kept a high standard for the awarding criteria. It should only be possible to mark a small percentage of marketed products with the Blue Angel. The label should only be awarded for a limited period and, if possible, the criteria should be strengthened in line with market penetration and further technological development to avoid a situation in which nearly the entire product market would be deluged with Blue Angels.

Phase 2 of the Blue Angel procedure starts with publication of Jury decisions through press releases from the BMU and the publication of the Basic Criteria document by RAL. Manufacturers are then able to apply for and conclude contracts on the use of the Blue Angel after the provision of all required information—for example, test reports of independent testing institutes, data sheets or other proof as specified in the Basic Criteria document—followed by a successful examination of the information by RAL, the experts of UBA and authorities of the federal states. For practical reasons, an examination of possible environmental problems deriving from the production facilities of the manufacturers does not apply to foreign applications.

With the conclusion of a contract with RAL, the manufacturer obtains the right to label the specific product with the Blue Angel and use it for advertising purposes. Depending on the annual turnover achieved with the product marked with the Blue Angel, the manufacturer must pay an annual fee to RAL: from a minimum of 178 euro for an annual turnover of up to 0.26 million euro to a maximum of 2034 euro for more than the 5 million euro. In addition, a single fee of 153 euro must be paid to RAL upon filing the application; an extra contribution of 20 percent of the annual fee must be paid into a fund administered by RAL and UBA for joint information campaigns. Compared to the costs of commercial advertisements, the financial burden for manufacturers seems to be negligible. To these costs, however, must be added the expenses for the required tests and verification data and these may significantly exceed the contribution for fees.

A last element of the awarding process concerns compliance with the criteria during the entire marketing period. The Blue Angel scheme does not provide for direct monitoring, but relies on control by competitors. Experience shows that this mechanism works very well. The German Comparative Testing Institute for consumer goods—Stiftung Warentest—examines consumer goods on a random basis and publishes the results in a monthly magazine. It has played a major role in ensuring the compliance of Blue Angel-marked products with the Basic Criteria. Up to now, no instance of noncompliance has been detected.

By January 2003, the requirements for approximately one hundred product categories had been defined, though some are no longer in use because either their objectives were achieved, they had been replaced by binding standards, or they were cancelled for other reasons. Manufacturers have signed contracts for the use of the Blue Angel in approximately 3,000 products, some 11 percent of which come from abroad. A large number of manufacturers—for example, manufacturers of computers and copying machines who have concluded contracts for the use of the label in the name of their German importers—are not reflected in this figure. (For the current status of the Blue Angel scheme, see http://www.blauer-engel.de.)

For a decade, the German environmental label remained the only one worldwide. It was not until 1988 that Canada and Japan started to develop their labeling programs; many other countries as well as the European Union followed in the 1990s. In 1999, nearly thirty countries made use of environmental labels (see OECD, 1999); all have been influenced to some degree by the Blue Angel program. Several so-called eco-labeling practitioners founded the Global Eco-labeling Network (GEN) in 1994 to foster information exchange and to represent environmental labeling schemes in international forums. Currently, twenty-five environmental labeling programs are members of the GEN. It supports ongoing international and national eco-labeling discussions through a variety of activities, such as providing data sources on criteria already defined for certain product categories. Up-to-date information is available on the Internet at the following Web site: http://www.gen.gr.jp. One of the objectives of GEN is to work towards a longer-term harmonization of national labeling schemes. The issue of harmonization and a discussion of its consistency with the purpose of the Blue Angel—that is, to foster competition and to provide an economic incentive to innovative forerunner companies—will be taken up in a later section.

Environmental labeling is on the agenda of international economic organizations such as the OECD and the World Trade Organization (WTO). In 1991, the OECD published an overview of existing and planned environmental labels in the OECD area (OECD, 1991), which contributed to a further diffusion of the instrument. In the second part of the 1990s, environmental labeling was discussed in the Committee on Trade and Environment (CTE) and the Committee on Technical Barriers to Trade (CTBT) of the WTO. The most controversial issue in the discussion of eco-labeling and trade is the consistency of environmental label schemes with the Agreement on Technical Barriers to

Trade (TBT). (For an overview of the issues that currently inform the discussions of eco-labeling and trade, see "Eco-labeling and the World Trade Organization" in OECD, 1999; see also Tietje, 1995; Piotrowski, & Kratz, 1999).

In summary, the Blue Angel is a very specific economic incentive tool and can serve as an example of an international diffusion of innovative environmental tools. It is not an instrument for raising general environmental awareness aimed at a fundamental change in consumer behavior. It is also not designed to promote directly research and development to solve problems not yet taken up by innovative forerunner companies. The Blue Angel tries to maintain a balance between ambition and high visibility. It follows a rather pragmatic approach that assumes, on the one hand, that overly ambitious criteria may lead to useless paperwork because no manufacturer will be able to apply and start the competitive process. On the other hand, it avoids setting criteria that are too weak, thereby perhaps increasing visibility but diminishing the economic value of the instrument over the longer term.

The Blue Angel is not an easy instrument to implement. A proper resource base is crucial for successful implementation. Programs that rely solely on self-financing run the risk of inflating, weakening, and finally devaluing the instrument. The Blue Angel is designed as an information tool. While the basic idea is simple, the underlying conditions, procedures and decisions are complicated and not easy to communicate.

Implementation problems

The Blue Angel had to overcome a number of barriers and meet serious challenges in order to make the market mechanism of supply (driven by the economic interests of manufacturers) and demand (driven by environmentally conscious consumers) work.

The high degree of environmental awareness in the German population helped launch the Blue Angel, though it did not necessarily shape the actual behavior of private consumers. Nonetheless, it had real impact since innovative manufacturers attached importance to this awareness. Decisive in promoting environmentally conscious behavior, however, was the role of trade and retailers.

During the first years, it was difficult to find labeled products in supermarkets and major stores. New products, along with those from small companies and not yet established brands, encountered difficulties finding space on store shelves. The marketing of Blue Angel products therefore initially started in specialized shops for major, more technical goods, whereas the day-to-day minor products such as deodorants or toilet paper had difficulty entering the market.

Realizing that the demand of private consumers would be limited, UBA during the early 1980s started to mobilize public procurement (UBA, 1981). In 1984, an ordinance was amended to allow the inclusion of environmental criteria in public tenders. In the following years, public authorities at the local, regional, and federal levels issued instructions for the public procurement services, obliging them to include environmentally friendly products in their tenders. In this regard, the Basic Criteria documents of the Blue Angel program were to specify the required quality of a certain product or service. Thus, the Blue Angel got some independence from the availability of Blue Angel marked products in shops. A survey conducted by UBA in 1997 found that in 86 percent of governmental institutions, guidelines, and recommendations for environmentally friendly procurement were in place (Umwelt, N. 7–8, 1997). The Blue Angel criteria, mainly those for office supplies and office machines such as copiers, are regularly introduced as a reference in public tenders.

From the perspective of manufacturers, the role of consumer demand in the beginning was a question of perception and of how much credit manufacturers would attach to the results of opinion polls on environmental awareness. It was something of a self-fulfilling prophecy and, for some product categories, it became reality when public demand emphasized green procurement strategies. In a number of cases, manufacturers also chose a niche policy. They continued to market their conventional products as well as environmentally improved ones of the same kind. According to market surveys, the group of especially environmentally aware consumers were people with rather high incomes who were expected to accept higher prices. Blue Angel marked products therefore were often more expensive than conventional goods. The price increase was not always justified because of higher production costs. Rather, it reflected the willingness of some consumers to pay more.

Media reporting on new Blue Angels was intensive and favorable for many years. It fed on an active public relations and press policy of the BMU and UBA. On the whole, the private media did most of the information work for the Blue Angel program. As early as the 1980s, surveys found that the Blue Angel was known by approximately 80 percent of German citizens. In the 1990s, from 60 to 78 percent of the interviewees indicated that they would prefer Blue Angel products to nonlabeled goods if they saw them on shop shelves (Neitzel, Landmann, & Pohl, 1995, p. 156f.). These figures have dropped in recent years, mainly among young people and the citizens of the eastern part of Germany where the Blue Angel is less known.

As we have seen, the Blue Angel process and procedures are based heavily on the broad participation of all relevant societal groups. The Blue Angel, however, is “an instrument that does not fit into cooperative structures based on consensus and the will of the majority" (BMU, 1990. p. 20). It expressly favors the interests of minorities and forerunner companies who in general have little weight and influence in the pluralistic institutional setup of German society. The experts of industry, trade, consumers, and unions in the jury and the hearings that had been appointed by their respective associations, therefore, have an unusual role to play. They do not have to represent and defend the interests of the majority of their clientele; rather, they make life for most of their clientele more difficult. This presented a dilemma and the way out was by no means a foregone conclusion. The history of the Blue Angel, therefore, has been very turbulent. It went through periods of obstruction, confrontation, and serious legal fights before a more cooperative relationship was reached with organized interests. That is, the Blue Angel program initially encountered almost unanimous resistance from industrial associations as well as the consumer community. While environmental groups had been favorable in the beginning, they too became increasingly critical as the Blue Angel advanced.

Role of organized industrial and commercial interests

To properly understand the role of the organized industrial and commercial interests, the general political and economic framework conditions should be recalled. The first oil crisis had adversely affected environmental policy by the time the Blue Angel program was launched in 1977. Industry successfully attacked environmental regulative policy and called for the use of flexible, market-based economic and voluntary instruments. The industrial lobby, therefore, did not openly oppose the new policy tool. Rather, it pursued a kind of hidden guerilla strategy. Industry gave lip service to the idea but refused to actively cooperate in the participatory forums and round tables created for the program.

During the first years the central representative of German industrial interests—the Federation of German Industry, or BDI—refused to nominate a member for the Jury and to nominate experts for the hearings. Given this obvious obstructionist strategy, UBA went public. It published work on criteria for specific product categories in the relevant journals of the affected industries and invited interested companies and manufacturers to address the BDI to appoint them as experts in the hearings. The strategy worked. The business community subsequently was represented by experts from forerunner companies as well as industrial lobbyists in the hearings; the economic interests of industry, in other words, were no longer represented as a monolithic block.

These were only preliminary skirmishes, however, in a more serious battle that commenced as the Blue Angel materialized.

As Blue Angel products began to appear on the market, the influence of the BDI became less relevant. Instead, heavyweights in the business community showed their muscle. Initially, they put economic pressure on individual companies. For example, the first Blue Angel categories were chlorofluorocarbon (CFC)-free sprays. The association of the German Chemical Industry (VCI) threatened a company, which filled deodorants and other cosmetics in cans and other marketable goods, that it would not get any new orders from the producers of these substances if it did not stop advertising its product with the Blue Angel. The company turned to the Federal Environmental Agency for help. The Agency forwarded the information to the authorities responsible for cartel control and abuse of economic power (Bundeskartellamt). Some months later, VCI was obliged to publish a statement in its official magazine indicating that VCI did not oppose advertisements for CFC-free products with the Blue Angel and acknowledging that the decision to do so remained entirely with individual companies (Müller, 1999a, p. 208).

Solidarity among German car manufacturers was much firmer. When the Jury started to include components of cars such as catalytic converters in the Blue Angel program, German car manufacturers— even those who exported cars equipped with catalytic converters to the U.S. market—declined to take advantage of this advertising opportunity. Moreover, they halted efforts at the staff level to prepare an application to use the label. It was a Japanese company that applied for the Blue Angel and started a huge advertising campaign for catalytic converter cars.

Another element of the German industrial establishment's confrontational strategy—one potentially very serious and dangerous—was to seek support in the courts. The association of the well established brands on the market (Markenartikelverband) was the primary instigator of litigation against companies using the Blue Angel in advertisements. It challenged the term environment friendly. The problem was that this litigation was never addressed to the originators of the Blue Angel criteria, the Jury or UBA. Rather, the exclusive purpose was to intimidate competitors, especially small companies. Some of the companies obeyed. They stopped using the Blue Angel. The more courageous ones faced legal challenges. They received support from the Federal Environmental Agency and from RAL and succeeded—sometimes only in the second instance—in keeping the Blue Angel and continuing an environmentally oriented marketing strategy (for an overview on legal problems see Janiszewski, 1992, p. 199ff). It was only in the late 1980s that the intimidation strategy more-or-less stopped and a more cooperative spirit prevailed in the relationship with strong economic actors.

Role of consumer associations

Much to the surprise of the initiators of the Blue Angel program, support from consumer associations was also difficult to obtain. The resistance of organized consumer interests, however, was of a different nature. Three elements played a role. First, the voluntary nature of the policy tool encountered skepticism. The state was viewed as the guarantor of environmental quality and product safety; this was to be assured through regulations. Second, institutional interests played a role. Consumer information as a voluntary mission was believed to be the responsibility of consumer organizations and they had, in fact, established a network of information centers in all German federal states. Since the 1960s, the Foundation of Comparative Testing of Goods (Stiftung Warentest) had published test results of common consumer goods such as household appliances and cleaners in its own publications available at any kiosk or bookshop. Third, the general concept of the Blue Angel and, in particular, the term environment friendly provoked criticism. The concern was that Blue Angel might mislead consumers. Instead of buying a bicycle, they would be encouraged to buy a cleaner running but still polluting car. Instead of reducing consumption, the Blue Angel might encourage consumers to consume more. Consumers would have a false sense of security.

With growing environmental awareness among consumers and increasing public debate on environmental matters, however, the attitude of consumer organizations changed rapidly. Again, the media played an important role in the process. Among the developments that reflected this change were the following:

- Consumer Centers actively included environmental aspects in their information work and used the Blue Angel, in the absence of other reliable tools, as a source of information to satisfy consumers' requests.
- In 1984, the Stiftung Warentest included the obligation to examine the environmental quality of goods in its statute and a representative from the Federal Environmental Agency became a member of the board of curators of the Stiftung Warentest.
- When the logo inscription environment friendly was changed to Environmental Label in 1988, the more ideological differences receded.
- Experts of consumer organizations now actively cooperate in the Jury as well as in hearings for the preparation of awarding cri-

teria. However, they still continue to keep a certain distance and are ready to object whenever they feel that environmental objectives dominate or even ignore consumers' interests concerning the serviceability and safety of a product.

Role of environmental associations

The third major partners important for the implementation of the Blue Angel program are the environmental associations. Given the diversity of the ecological movement in Germany and the competition among the different groups, they played and continue to play an ambivalent role. From the very beginning, representatives of environmental organizations cooperated actively and successfully in the Jury. They helped generate public trust in the experts of UBA and supported strongly their noninflation strategy. These representatives, however, have acted more or less in their own personal capacity. Consequently, they have not been able to serve as effective transmission belts to their organizations, thus limiting their ability to mobilize support.

The publicly articulated position of organized environmentalists over the many years has been a rather critical one. Similar to consumer associations, they rejected the term environment friendly and wanted to see a more fundamental influence of the Blue Angel program on consumer behavior. In many cases the awarding criteria were heavily criticized for being one-dimensional, not taking into account the entire life cycle of a product. Blue Angel criteria, they argued, should reflect not only the state of the art, but rather, awarding criteria should be used like blueprints for manufacturers to stimulate further progress and environmental improvements of products, even if no domestic or foreign manufacturer would be able to apply for the use of the label at that time.

Cancellation of the expression environment friendly also has satisfied environmentalists. Nonetheless, though it seems that the German ecological movement in general accepts Blue Angel as a market-oriented tool of environmental policy, it does little to promote the purchase of products marked with the Blue Angel.

In sum, given the initial acceptance and support of the program by powerful actors and groups was rather problematic, in many ways it is surprising that Blue Angel survived. That the Blue Angel program has survived and advanced can be attributed to the support of three allies: (1) private and, in particular, public consumers (i.e., the public procurement policy); (2) innovative individual manufacturers, domestic and foreign, that seized the opportunity to increase their market share; and (3) the media and journalists who, by reporting about successes and barriers, helped make Blue Angel known to the general public.

Two additional factors in the survival of Blue Angel, however, were equally important: the solid institutional set up and strong political support for the program. Throughout its history, all German environment ministers have had an interest in the program. It has enjoyed the continuous support of the staff of UBA and the experience of RAL as certifying institute, as well as a high degree of continuity in the composition of the Jury. All of these factors have been assets.

All told, the Blue Angel has grown and matured over time. The economic and ecological impact of Blue Angel on innovation, and its role in the toolbox of environmental policy are discussed below.

Innovation and the role of environmental labeling in the environmental policy toolbox

Research on environmental policy that fosters innovation identifies a number of elements important to successful policy implementation (Blazejczak, Edler, Hemmelskamp, & Jänicke, 1999, p. 15). According to this work, policies create a friendly environment for innovation if they follow a strategy which:

- uses instruments that provide economic incentives, combines several instruments, is based on strategic planning, target setting and takes into account the process and different phases of innovation.
- applies a policy-style that favors dialogue and consensus, is reliable, sustainable, continuous, proactive, demanding, flexible and management oriented.
- creates a network of public actors, private actors and stakeholders and promotes interaction between them.

The Blue Angel program corresponds to several of these preconditions, mainly with respect to the policy style and established network. It shows that reliability, continuity, and proactive, demanding management are even more important than reaching a consensus among all organized stakeholders. A strong role for public actors, therefore, is important for a successful and efficient working network (see also Voelzkow, 1996, p. 311). The weak position of forerunner companies relative to the organized commercial interests calls for a neutral, fair and target-oriented management that seems to be best provided by public authorities. The following discussion focuses on the economic and ecological effectiveness of the instrument and its impact on innovation and diffusion.

Strategic planning or incremental policy-making?

Strategic thinking and long-term, comprehensive planning led to the inclusion of environmental labeling in the toolbox of the first German environmental policy program. In the process of putting the program into action, however, there was no systematic, long-term strategic planning concerning the integration of the tool with other policy instruments. When the Blue Angel program was launched, a learning by doing process started. This resulted in a pragmatic search for supporting tools that could help foster the Blue Angel mission. The development of the Blue Angel program, therefore, shows more elements of incremental policymaking than strategic planning.

To recall a point made earlier, the German Blue Angel was the child of crisis. It was launched at a time of economic recession when environmental policy did not have an easy time. Former German Minister for the Environment, Klaus Töpfer put it correctly when he said: "A lack of political power to implement measures had to be compensated for by sheer creativity" (BMU, 1990, p. 17). Environmental regulative policy had been criticized as a barrier to innovation, new investment and job-creation. Environmental policymakers had to convince all those critical domestic political and economic powers that innovation could be stimulated while at the same time protect the environment by the use of information and voluntary economic incentives. Development of the internal European market and the policy of the European Communities (EC) reenforced the need for new initiatives in product policy. It was increasingly impossible to set product standards unilaterally. Otherwise, the chance for an agreement among EC member states on sufficiently ambitious environmental product standards harmonized at the European level was minimal at that time. Environmental policymakers, therefore, sought to make use of their main allies: the high degree of environmental awareness voiced by the general public and state-of-the-art environmental technology that had been actively promoted through public subsidies for research and development.

Market impact of the Blue Angel Program

There have been several attempts to evaluate the market impact of the Blue Angel Program and its impact on marketing strategies of companies (OECD, 1997; UBA, 1998). The findings show a mixed picture:

- A major impact could be mainly observed in those product categories that are interesting for professional purchasers and public procurement; however, the market share of Blue Angel paints was above 60 percent in the do-it-yourself-sector, but only 20 percent in the handicraft sector (figures for 1995, OECD, 1997, p. 50).
- Label-using companies are more or less equally represented among small, medium, and large companies (UBA, 1998, p. 11f).
- Microeconomic benefits of individual forerunner companies disappear overtime as other companies enter the market with environmentally improved products. Smaller companies reported the greatest impact on their market position, whereas medium sized companies were among those that often observed little or no effect on their market share (UBA, 1998, pp. 35, 36).
- Price management is an important factor for the marketing strategies of companies; however, price elasticity had declined over the period the Blue Angel has been operating. The assessment study of UBA detected a decline in the willingness of German consumers to pay more for the environment in the late 1990s compared to the 1980s and the early 1990s (UBA, 1998, p. 39).

In recent years, the number of environmental product labels has increased dramatically. Another interesting result of the assessment study of the Federal Environmental Agency is that the multiplication of environmental labels has not reduced but rather increased companies' interest in the Blue Angel. The reason is its great credibility in the eyes of consumers and the fact that it is well known (UBA, 1998, p. 17).

Impact of the Blue Angel Program on innovation and diffusion

The contribution of the Blue Angel to mitigate or solve product-related environmental problems by stimulating innovation and diffusion shows a mixed picture, too. Innovation in this context means all changes in the life cyle of a product—major or small—resulting in a reduction of pollution. Diffusion addresses the degree and rate of market penetration of the improved products. Let me illustrate this with some examples.

Among the first Blue Angel products were returnable bottles. Waste policy desperately looked for ways to stop the decrease of return bottles used for soft drinks, milk, and beer. For this kind of problem the Blue Angel was a total failure. The Basic Criteria for returnable bottles have been improved over the years by adding requirements forbidding the use of lead-containing bottle caps and labels containing heavy metals. Forty-one companies for sixty-eight products used the label in advertisements in 1998, but they could not stop the overall increase of one-way bottles and other packages for beverages. Environmental policy looked for other instruments. In the early 1990s, the ordinance on package material was introduced together with the dual system and the Grüner Punkt for packages. The ordinance set the threshold of a 72 percent rate of returnable bottles used for soft drinks; otherwise, an obligatory deposit on one-way bottles and cans should be imposed. Against strong opposition of retailers the deposit was finally introduced in 2003.

Another example of the Blue Angel's limited impact was its adoption for CFC-free sprays in 1978. Despite a number of companies that used the label, Blue Angel could not increase significantly sales for alternative substances and mechanical devices used in sprays. It did, however, alert the chemical industry to the need to develop less environmentally harmful solutions if the market for sprays and cosmetics was not to be lost. As a result of the Montreal protocol on ozone depleting substances, the production and use of CFCs is now banned.

Products made of recycled paper such as toilet paper and kitchen rolls have been promoted successfully by the Blue Angel. Recycled paper for graphics paper, copying paper, envelopes, dispatch bags, calendars and similar uses also were promoted successfully by the Blue Angel. According to information from the German association of paper mills, the percentage of recycled paper used in the entire production of all kinds of paper products was 54 percent in 1993 (Neitzel, Lundmann, & Pohl, 1995, p. 160). The demand was not driven by private consumers but by public procurement policy. Public demand, in general, actively promoted all product categories related to office supply and office appliances such as copiers, printers, and computers.

The Blue Angel awarded to low-noise construction machinery such as wheel loaders, road-making vehicles, concrete mixers and motor compressors was also very successful. Within a very short period, the more expensive products with the Blue Angel label could significantly increase their market share. The success was supported by officially introduced user benefits: these machines were allowed to work in sensitive areas such as near hospitals or during times when regulations forbade noisy work.

The Blue Angel for heating appliances such as low-emission oil and gas burners is another successful, but different, example of the effect of this kind of policy tool. Supported by providers of gas (mainly Ruhrgas) and accompanied by training courses organized by craft organizations, environmentally improved burners rapidly penetrated the market. The awarding criteria could be strengthened several times. This technological progress also had an effect on regulation, resulting in a strengthening of the legally binding emission standards.

These examples show that the ecological effectiveness of voluntary environmental labeling and its usefulness for overall environmental policy depends very much on the nature of a product category. Products that are part of day-to-day consumption and purchasing of private consumers are the most difficult. Given the rather passive role of retailers in the use of the Blue Angel, it is understandable that consumers' response is limited if the consumer has to invest time and effort to purchase, for example, environmentally improved toilet paper, because he or she does not immediately find the labeled product in the local supermarket.

Another bottleneck in the market of private household consumption is the declining interest of companies in those product categories where environmental performance over the past decade improved dramatically and where environmental claims in advertising campaigns are widely used. This is the case for detergents and household equipment such as washing machines, dish washers, stoves, and refrigerators. There are several reasons for limiting the use of the Blue Angel as an incentive for promoting further innovation. Most important may be the well developed technical level of these products and the fact that all major companies work more or less in parallel on further improvements. Thus, no single company hopes for benefits from a first mover advantage. An additional factor may be that all providers offer a variety of models on the market. As a consequence, they hesitate to advertise special, better models since this may adversely impact the rest of their product line.

Despite these limits, the Blue Angel has successfully promoted innovation in a number of product categories. In the case of varnishes and burners especially, it stimulated continuous efforts for further improvements and innovation. This was accompanied by rapid diffusion and market penetration. An indicator of success is the stagnation of the number of products labeled with the Blue Angel. Over the past decade, it has remained around 4,000 each year. On average, some 500 to 800 expired contracts were replaced by roughly the same number of new contracts; among those contracts were product categories with further improved and strengthened criteria (see http://www.blauer-engel.de). The Blue Angel program seems to be fully in accordance with the conclusions of an OECD study. It claimed that the environmental benefit of eco-labeling "will be achieved when a balance is reached between the number of eco-labeled products and the stringency of the criteria" (OECD, 1997, p. 7). A large number of certified products alone is no indicator of success. Continuously improved criteria may be a precondition for eco-labeling to achieve positive environmental effects.

Blue Angel: Useful part of a policy mix

The role of the Blue Angel in view of the overall toolbox of environmental policy shows two different faces. On the one hand, the impact of the Blue Angel was improved significantly when it was supported by and complemented with additional tools and measures. On the other hand, the Blue Angel was often a door opener and facilitator in preparation for binding regulation. Let me elaborate.

Green public procurement policy was a major driver for the Blue Angel. It was decisive for the success of all product categories of relevance in the public sector. The awarding criteria progressively developed towards technical guidelines which can be introduced in private and public tenders independently from the physically purchasable product. The president of UBA sees here the future of eco-labeling. He said on the occasion of the twentieth anniversary of the Blue Angel program: "[Ecolabeling criteria should] provide all companies with guidance on how the product should be improved and what requirements they could be expected to satisfy as soon as tomorrow. . . . it is allowed to claim that within the framework of eco-labeling ecological standards for products are created to supplement the technical standards developed by standardization bodies such as ISO or CEN for Europe" (OECD 1999, p. 16). It is obvious the impact of environmental labeling on innovation could be significantly strengthened if the observed trend in some product categories could be broadened.

Other tools such as user benefits in regulations strongly supported such successes as the Blue Angel for construction machinery. Furthermore, the Blue Angel program, in several cases, profited from the fact that experts of UBA involved in Blue Angel work were also in charge of research and development projects. They seized the opportunity to promote the implementation of the results of their projects through the use of Blue Angel.

Blue Angel is also an interesting tool to prepare a national economy for future demands and obligations that may result from national legislation or binding international agreements. The Blue Angel for CFC-free deodorants, for example, was not very effective in reducing ozone-depleting emissions; however, it did stimulate the awareness of producers and consumers. When the use and production of CFCs was finally banned by regulation, the chemical industry had already developed substitutes and therefore had no problems meeting the obligations of the Montreal protocol and the related European and national legislation.

The lessons from the Blue Angel program show that voluntary environmental labeling schemes are also interesting in view of the incremental nature of environmental policy-making. The Blue Angel initiated and continues to initiate a trial and error process. The results help to better assess political measures. If the soft tool works—and I showed that it does so in some cases—it is worthwhile to invest further efforts and resources in its improvement. If it does not work, it provides arguments and scope for action to introduce more adequate measures to address environmental problems. Again, in several cases the Blue Angel proved to be an excellent door opener for more stringent measures.

The described potential and effectiveness of environmental labeling raises trade related concerns. As mentioned, there are ongoing discussions about possible trade barriers, not only for trading among industrialized countries, but also for trade opportunities of developing countries (Piotrowski & Kratz, 1999). These discussions primarily address the production phase and whether environmental labeling should define standards independently from the legal situation and state of technology in different countries. Consensus could be reached on the need to distinguish between product-related production and process methods, the so-called PPMs and overall requirements for plants. The Blue Angel approach may be an example of how to reconcile these trade concerns. It excludes those PPM-related criteria that cannot be identified in the product itself, but it includes the production phase whenever it is directly relevant for the environmental impact of a product. Examples are paper and chemical products such as paints and varnishes, or products and services for nontoxic indoor pest control and prevention.

Some defenders of international free trade request that national environmental labeling schemes should either be harmonized or an agreement on international guidelines on equivalence and mutual recognition should be reached to avoid trade barriers (UNCTAD, 1994, 1995). As a policy tool, environmental labeling can, however, only reap positive results if its specifics are recognized. The worldwide or even European-wide harmonization of environmental labels would destroy the economic incentive character of the tool for innovative forerunner companies. The political debate on the harmonization of environmental labels, to my mind, shows a lack of understanding of the importance of the political level on which policy tools can successfully operate (Müller, 1999b, p. 137). A harmonized environmental labeling scheme would be welcome if there were a chance to maintain the specific character of the tool—that is, to stimulate competition on environmental performance and quality. Environmental labeling schemes are voluntary tools; they are designed to empower forerunner companies. Very often, these companies are in an outsider, minority position and have difficulties successfully placing their products in a market dominated by strong and established economic actors. Environmental labeling will not restrict competition as long as the requirements for the use of the label are transparent and labels are accessible to all domestic and foreign companies able to meet them.

Comparative empirical studies on different national environmental labeling schemes, their successes and problems (Harrison, 1999; Eiderstroem, 1998; Frey & Iraldo, 1999) clearly indicate that the progress and effectiveness of environmental labels depend on the given situation in national product markets, the level of technological progress, specific national economic interests and the environmental consciousness of consumers. For example, a harmonized European label that would take a rather low level of environmental quality of products in some EU-countries as a reference, would inflate the label in other countries where the usual standard is already higher. As a consequence, it would be of no value in those countries. On the other hand, if it would take the highest technological level in EU-member states as a reference, consensus would be difficult to reach. The EU-Commission therefore has dropped its previous demand that national and regional environmental labels be discontinued (Piotrowski & Kratz, 1999, p. 437). To my mind, environmental labeling schemes will only be a useful complement to the environmental policy toolbox if it is used in a subsidiary and competitive manner. It has to be designed according to specific problems and specific conditions that vary from country to country and from market to market.

This conclusion does not exclude international cooperation and competition. To the contrary, lessons from the Blue Angel program show the importance of international competition. Some Blue Angel categories only progressed because foreign companies applied. On this signal, national companies decided to follow. In addition, efforts to improve information exchange on criteria and progress of national labeling schemes are most welcome, not least because they facilitate the work of national experts to define demanding new—and revise already existing—awarding criteria. The nature of voluntary national environmental labeling programs calls for competition among labeling schemes. "That is why eco-labels need public reporting, need the media. . . . [I]n the medium term the better and more ambitious labels will have prevailed" (President of the Federal Environmental Agency, in: OECD, 1999, p. 19f).

I also would like to correct what, in my view, is another frequent misperception regarding the efficiency of certain environmental policy tools. Academic publications of economists have contributed to the false perception of voluntary soft instruments. State regulation often has been stigmatized as being inefficient, whereas voluntary and, in particular, economic instruments have been praised not only because of their presumed high level of effectiveness but also their efficiency in terms of resources needed (Müller, 1999b). The example of the Blue Angel shows that the reality is much more complex. Depending on a number of framework conditions, voluntary economic tools can be effective but failures in meeting their objectives are also possible. Concerning efficiency, there is certainly not much difference with other, in particular, regulative tools. The proper management of the tool is resource intensive. Sufficient human, financial, and time resources are needed to develop, implement, and promote the tool. Without the support of a broad range of specialists of the Federal Environmental Agency, the work would certainly not have been done. Entirely self-financing schemes that depend on fees and contributions of the label users will either have difficulty advancing (as is the case of the Euromargerite, the European environmental label) or they will have difficulty keeping a high, demanding level for the criteria resulting in a smaller number of awarded labels and minor revenues from fees. Sufficient financial resources are also needed for public relations activities. The label does not necessarily sell its information by being present and visible on the shopping shelves. Independent information campaigns have to compensate for the lack of high visibility. Despite a rather limited budget, the Blue Angel profited from the gratis reporting of journalists and joint PR activities financed from the RAL fund, which is financed from fees paid by the label users. This fund, however, was never large enough to launch broad advertisement and information campaigns. It might have been possible to avoid some of the problems of the Blue Angel program and increase its effectiveness if more financial resources had been made available to intensify the information work.

Conclusion

The Blue Angel has helped promote innovation in some cases and has significantly promoted the diffusion of best available technology to reduce product-related environmental problems in quite a number of product categories. This happened primarily when it was accompanied by additional tools and when, in the view of industry, regulatory measures were in the political pipeline. "The environmental label is just one instrument of environmental policy" (Klaus Töpfer, former German Minister for the Environment, OECD, 1991, p. 31). The lessons drawn from the Blue Angel program, therefore, are identical with the conclusion of an empirical study on environmental labeling schemes in Canada, the European Union, and the Nordic Countries carried out by Harrison. It reads: "Eco-labeling programs are . . . best viewed as a complement, rather than an alternative, to a broad range of governing instruments . . ." (Harrison, 1999, p. 34).

Environmental labeling programs, however, should not be underestimated. Environmental policy does not have at its disposal many influential tools to promote innovation and diffusion of environmentally improved products. Seen from this perspective, voluntary environmental labeling schemes are very useful and should be part of a well assorted environmental toolbox. From the Blue Angel viewpoint, its impact on innovation could be further improved by better collaboration with public research and development (R & D) programs. More systematic cooperation between environmental R & D programs and environmental labeling activities would offer even better opportunities for both to become more effective. The effective use of important public R & D funds is not always certain, because the responsibility of the research and development fund providers ends where implementation efforts should start. Voluntary environmental labeling schemes are perfectly designed to take over whenever research and development results are ready for marketable products.

Environmental labeling can help to accelerate diffusion of technical improvements of products. They are not in a position to promote social and fundamental changes of consumer behavior. If they would intend to do so, the economic incentive nature of the tool for manufacturers would be lost. The use of bicycles, for example, must be promoted primarily by the construction of safe roads and paths for bicyclists. A label cannot do this job. For environmentally conscious consumers who otherwise would feel lost and frustrated when confronted with messages about environmental threats and future catastrophes, the environmental label offers opportunities for concrete, responsible action.

Research on environmental awareness and environmentally conscious behavior of German citizens show encouraging results. While opinion polls point to a diminishing rate of expressed environmental awareness, the concrete behavior of German citizens does not show any change. A majority of Germans continue to actively participate in all kinds of recycling activities and make rather environmentally conscious consumer choices, if these choices are not too cumbersome. Certain attitudes have been stabilized and seem to be part of an unquestioned routine (Preisendörfer, 1999, p. 71, Neitzel, Landmann, & Pohl, 1995, p. 167ff).

To conclude, the lessons of the Blue Angel program support the findings of environmental policy research that call for a sophisticated toolbox for environmental action. Environmental labeling programs are not less complex or more efficient in terms of consensus and resources needed than other environmental policy tools. Attention should be concentrated on choosing a mixture and assortment of different types of policy tools in accordance with the nature of the problems to be solved. The choice of the right tool is largely dependent on the political, technological, and societal framework conditions and the political level at which decisions have to be taken and production and consumption takes place. Environmental labels should not be a subject of international harmonization. Instead, different national labeling schemes should compete with each other. Those with weaker criteria will either improve or disappear. Winners will be innovative forerunner companies and thus the environment.

References

Blazejczak, J., Edler, D., Hemmelskamp, J., & Jänicke, M. (1999). Umweltpolitik und Innovationswirkungen im internationalen Vergleich. *Zeitschrift für Umweltpolitik & Umweltrecht, 1*(99), 1–32.

BMU/Federal Minister for the Environment. (1990). *International conference on environmental labeling.* State of Affairs and Future Perspectives for Environment Related Product Labeling, July 5–6, 1990 in the Reichstag, Berlin.

Deutsches Institut für Gütesicherung und Kennzeichnung e.V. (RAL) (2000). *Umweltzeichen. Environmental Label. German "Blue Angel."* Product Requirements, Edition July 2000. Sankt Augustin.

Eiderstroem, E. (1998). Ecolabels in EU Environmental Policy. In J. Golub (Ed.), *New instruments for environmental policy in the EU.* (pp. 190–214). London and New York: Routledge.

Federal Environmental Agency (UBA). (1998). *Assessing the success of the German eco-label.* Texte 61/98, Berlin.

Frey, M., & Iraldo, F. (1999). *Promoting Eco-Labeling Schemes. Promotion and diffusion of the Eco-label: a "network" model application in the EU.* OECD, Conclusions and Papers Presented at the International Conference: Green Goods V <Eco-Labeling for a Sustainable Future>, Berlin Conference, October 26–28, 1998, ENV/EPOC/PPC(99)4/FINAL: 98–114.

Harrison, K. (1999). Racing to the top or the bottom: Ecolabeling of paper products in Canada, Scandinavia, and Europe. *Environmental Politics, 8*(4), 110–136.

Jänicke, M., & Weidner, H. (Eds.). (1995). *Successful environmental policy. A critical evaluation of 24 cases.* Berlin.

Janiszewski, J. (1992). *Das Umweltzeichen: rechtliche Analyse und Neuorientierung.* Berlin.

Kemp, R. (1997). *Environmental policy and technical change. A comparison of the technological impact of policy instruments, (New horizons in environmental economics)*. Edward Elgar Pub.: Cheltenham, UK & Brookfield, Vt.

Kern, K., Jörgens, H., & Jänicke, M. (1999). Die Diffusion umweltpolitischer Innovationen. Ein Beitrag zur Globalisierung von Umweltpolitik. FFU-Report 99-11, Berlin.

Müller, E. (1986). *Innenwelt der Umweltpolitik. Sozial-liberale Umweltpolitik—(Ohn)Macht durch Organisation?* Opladen Westdeutscher Verlag.

Müller, E. (1997). Handwerkskasten der Umweltpolitik. In F. Biermann, S. Büttner, & C. Helms (Eds.), *Zukunftsfähige Entwicklung. Herausforderungen an Wissenschaft und Politik* (pp. 256–272). Berlin Edition Signa.

Müller, E. (1999a). 25 Jahre Umweltbundesamt—Spuren in der Umweltpolitik. Zweiter Teil. *Jahrbuch Ökologie 2000* (pp. 199–220). München: Beck'sche Reihe.

Müller, E. (1999b). Ökonomische Effizienz und politische Effizienz in der Umweltpolitik. In E. Gawel & G. Lübbe-Wolff (Eds.), *Rationale Umweltpolitik—Rationales Umweltrecht. Konzepte, Kriterien und Grenzen rationaler Steuerung im Umweltschutz* (pp. 119–139). Baden-Baden: Nomos Verlagsgesellschaft.

Neitzel, H., Landmann, U., & Pohl, M. (1995). Zur Empirie der Sustainable consumption, (Verantwortlicher Konsum): das Umweltverhalten der Verbraucher—Entwicklungen und Tendenzen—Elemente einer, Ökobilanz Haushalte. In B. Seel, C. Stahmer (Eds.), *Haushaltsproduktion und Umweltbelastung: Ansätze einer Ökobilanzierung für den privaten Konsum* (pp. 126–176). Frankfurt/Main Stiftung der Private Haushalt.

Organization for Economic Cooperation and Development (OECD). (1991). *Environmental labeling in OECD countries*. Paris: Author.

Organization for Economic Cooperation and Development (OECD). (1997). Case study on eco-labelling schemes. Presented at the Joint Session of Trade and Environment Experts, April 28-29, 1997, COM/ENV/TD(96)69/FINAL. Paris: Author.

Organization for Economic Cooperation and Development (OECD). (1999). *Conclusions and papers presented at the international conference: Green Goods V <Eco-Labelling for a Sustainable Future>*. Berlin Conference October 26–28, 1998, unclassified document, ENV/EPOC/PPC(99)4/FINAL. Paris: Author.

Piotrowski, R., & Kratz, S. (1999). Eco-labelling in the globalised economy. *Internationale Politik und Gesellschaft* (4), 430–443.

Preisendörfer, P. (1999). *Umwelteinstellungen und Umweltverhalten in Deutschland: empirische Befunde und Analysen auf der Grundlage der Bevölkerungsumfragen "Umweltbewußtsein in Deutschland 1991–1998."* Umweltbundesamt, Opladen.

Rat von Sachverständigen für Umweltfragen (SRU). (1978). *Umweltgutachten 1978*. Deutscher Bundestag, Drucksache 8/1938.

Risse-Kappen, T. (1994). Ideas do not float freely: Transnational coalitions, domestic structures, and the end of the Cold War. *International Organizations, 48*(2), 185–214.

Tietje, C. (1995). Voluntary eco-labelling programmes and questions of state responsibility in the WTO-GATT legal system. *Journal of World Trade: Law, Economics, Public Policy, 29*, 123–158.

Umwelt. (1997). Bundesministerium für Umwelt, Naturschutz und Reaktorsicherheit, ed., N.7–8, Bonn.

Umweltbundesamt. (1981). *Umweltschutz in der öffentlichen Vergabepolitik.* Texte 3/81, Berlin.

Umweltprogramm der Bundesregierung. (1972). *Das Umweltprogramm der Bundesregierung.* Mit einer Einführung von Hans-Dietrich Genscher, Stuttgart et al.

United Nations. (1992). *Agenda 21: A blueprint for action for global sustainable development into the 21st century.* New York: Author.

United Nations Conference on Trade and Development (UNCTAD, October 6). (1994). *Eco-labelling and market opportunities for environmentally friendly products.* Report by the UNCTAD Secretariat, TD/B/WG.6/2.

United Nations Conference on Trade and Development (UNCTAD). (1995, March 28). *Trade, environment and development aspects of establishing and operating eco-labelling programmes.* Report by the UNCTAD Secretariat, TD/B/WG.6/5.

Voelzkow, H. (1996). *Private Regierungen in der Techniksteuerung. Eine sozialwissenschaftliche Analyse der technischen Normung.* Frankfurt, New York Campus.

World Summit on Sustainable Development. (2002, September 4). *Political declaration and Johannesburg plan of implementation* (advanced unedited text).

CHAPTER THREE

Global Civil Society and Global Environmental Protection

Private Initiatives and Public Goods

Ronnie D. Lipschutz

Introduction

A critical set of issues arising from contemporary globalization are organizational, social, economic, and environmental externalities accompanying activities undertaken across national borders (Lipschutz, 2002). In the interests of economic competitiveness and growth, nation-states have yielded or restructured a substantial amount of their domestic regulatory authority to transnational regimes and organizations, such as the World Trade Organizatioin (WTO), the International Monetary Fund (IMF), and various other international regimes and institutions (Strange, 1996; Mishra, 1999). While globalization is much discussed in terms of the mobility of capital and production, its disruptive effects on environmental protection are as important, if not more so (Lipschutz, forthcoming).

While the conventional approach to regulation of transborder environmental externalities has been through treaties, conventions, and international regimes, in the face of a number of international regulatory bottlenecks, growing attention is being paid to the role of business and industry in generating these problems. As a result, regulation is increasingly the product of private (as opposed to public) interventions into global trade, corporate behavior, and consumer preferences. In the face of an international failure to establish a global forestry convention (Lipschutz, 2001), such initiatives have proliferated, offering competing venues for those interested in fostering sustainable forestry (Meidinger, 2003).

This chapter represents an attempt to explain, categorize, and describe such institutional developments in the environmental issue area, with a particular focus on nongovernmental organizations and global civil society (Lipschutz, 1996) and their role in environmental governance and regulation in the area of sustainable forestry management (SFM). The first part of this chapter briefly discusses the changing global regulatory environment and some of its causes. The second section explains the primary tools used in private SFM initiatives, including certification and eco-labeling. The third section catalogues and describes several of these private and semiprivate regulatory regimes. Finally, these initiatives are assessed in light of the framework presented by the volume's editor in chapter 1.

The new international division of regulation

Why regulate? Sometimes, there is judged to be a need to constrain the activities of producers in order to maintain competitiveness and efficiency in markets and to avoid the social costs of monopoly. Regulation is frequently judged necessary to reduce or eliminate externalities, both environmental and social, which otherwise provide undeserved benefits to producers. These and other forms of regulation have historically emerged through institutionalized political processes within states, especially when it has become glaringly apparent that self-regulation is nonexistent or inadequate. Whether regulations are too lax or too heavy, or what their particular form takes, is not at issue here; it is the fact that within states there exist mechanisms to regulate externalities that is important. Moreover, the ability and right to demand such controls, have them implemented, and achieve some degree of distributive justice is critical to system legitimacy.

By contrast, there is no standardized procedure whereby such regulations might be promulgated internationally. To be sure, rules and regulations are being formulated and implemented all the time, through a broad range of international organizations, regimes, and agencies, but each rulemaker or rule-making forum does so in a fairly idiosyncratic fashion and rarely with consideration of or in consultation with others. Some of these rules and regulations have the force of international law, and are meant to be implemented through domestic legislation and enforced by domestic authorities. Others are administrative tools, whose application is primarily functional and sectorally-limited (as in the case, for example, of commercial aviation, telecommunications frequencies, or geosynchronous satellite slots). A third category involves limits or prohibitions on certain types of national activities or legislation (resulting, for instance, from the dispute resolution process of the WTO or the rather weak oversight of the International Atomic Energy Agency).

International regulation has not always been as public an affair as it is today (and James Scott argues that, even today, much regulation is customary rather than public; see Scott, 1998, chap. 1). Historically, major social activities within societies were governed by customs, laws, covenants and contracts among and between individuals and groups, often but not always with the approval or support of some legitimate public authority (Braithwaite & Drahos, 2000). For example, medieval guilds formulated strict rules governing membership and practice, and this form of self-regulation has been carried into the medical and legal professions (which, nevertheless, are permitted to regulate their members only with the explicit authorization of local, provincial, and national governments). Maritime law is an arena where there has long been and continues to be a considerable amount of private regulation (Cutler, 1999). Another example can be found in common pool resource systems, such as those described by Elinor Ostrom (1990) and others (Bromley, 1992).

After World War II, most regulation remained national and state-sanctioned. There were certain sectors in which international public regulation was instituted, as in the control of the spread of nuclear weapons, the allocation of radio and television frequencies and geosynchronous satellite slots, and so on (Haas, 1992). There were, as well, private organizations that certified the quality and performance of other private organizations, such as the Better Business Bureau, *Good Housekeeping*, the Consumers' Union in the United States and the Consumers' Association in the United Kingdom. In a few cases, national regulatory systems were internationalized and recognized as the basis for regimes. For example, the safety rules of the U.S. Federal Aviation Administration (FAA) have been generally adopted by all national aviation authorities, although they are not always rigorously followed. Domestic public regulation also had the concomitant effect of limiting entry into markets and professions, and international regulation has had much the same effect in areas such as nuclear weapons development and agricultural trade.

Today, the state monopoly over regulation appears to be well past its twentieth century apogee. With the onset of neoliberal globalization, two things have happened. First, externalities once restricted to domestic jurisdictions and national oversight have burst the bounds and boundaries of the nation-state (Lipschutz, forthcoming). Second, no commensurate and comprehensive system of international regulation has been established to control these externalities. Why is this? The reasons are to be found in what is often claimed—an assertion contested by some—to be the relentless pressures of market competition and its efficient results when left underregulated, rather than the absolute diminution of the state's power or the supposed international anarchy that exists among states. In particular, in the interests of economic competitiveness and growth, states have been decentralizing, deregulating, and liberalizing so as to provide more attractive economic environments for financial capital (Strange 1996; Mishra, 1999; for an argument that such deregulation has not taken place, see Vogel, 1996). Strictly speaking, this form of deregulation is not a new phenomenon; it began in earnest after World War II, with the Bretton Woods institutions (Ruggie, 1983; Lipschutz, 2000, chap. 2). What is new today is the scale on which domestic regulatory authority is being transferred to international regulatory regimes and organizations, such as the WTO, and shifted from command-and-control regulation to what can be called surveillance and market-based regulation (Lipschutz, 2000, chap. 7). One result is that there is a growing tendency by some governments to implement policies attuned to a global economy through regulatory harmonization. This is especially true in Europe, where future candidates for membership in the European Union (EU) are required to write the Union's environmental and social provisions into new legislation (Wiener, 1999).

At the global level, however, such harmonization has been restricted to certain sectors, especially those having to do with trade and, to a lesser degree, the environment. Social regulations are pointedly excluded (Braithwaite & Drahos, 2000; Zaelke, Orbuch & Houseman, 1993) and discouraged by international financial institutions such as the IMF. Consequently, as governments have proceeded along the path to liberalization, and have sought to reduce social costs to corporations in order to attract foreign investment, the safety net historically provided by the welfare state has been degraded, with social responsibilities either abandoned or yielded to private actors and nonstate organizations (Mishra, 1999). That safety net, it should be noted, includes not only guarantees of health and safety, environmental protection, public education, and so on, but also standard sets of rules that level the economic playing field and ensure the sanctity of contracts. The latter two are especially important to capital, which demands both political stability and, if possible, harmonization of regulations in order to minimize transaction costs and investment risk.

As states have shed these responsibilities, some are being taken up by other institutions. A growing number are private or semipublic, some are transnational, others local or regional, and all are organized by civil society actors or corporate organizations. As Errol Meidinger has observed, writing about environmental regulation:

> Private organizations have recently established numerous programs aimed at improving the environmental performance of industry. Many of the new programs seek to define and enforce standards for environmental management, and to make it difficult for producers not to participate in them. They claim, explicitly and implicitly, to promote the public interest. They take on functions generally performed by government regulatory programs, and may change or even displace such programs. Private environmental regulatory programs thus have the potential to significantly reshape domestic and international policy institutions by changing the locus, dynamics, and substance of policy making. (Meidinger, 1999–2000, p. 2)

The forestry sector offers an especially apposite illustration of these claims because, as we shall see below, it is one in which a large number of private regulatory programs have emerged.

What's at stake in the Sustainable Forestry Management debate?

Why regulate forestry practices at all? Aside from the intrinsic ecological value of the various species of trees themselves, forests serve a variety of ecological roles, providing habitat for other plant and animal species, environmental services such as water purification, soil retention, local climate moderation, and carbon sequestration (with the last being especially important for global climate), and as reservoirs of genetic diversity. These services are not provided in equal terms by forests managed purely for growth, and the rate of destruction of non-managed forests, especially in tropical regions is, by all accounts very high. There is, in other words, a global public interest in seeing that forests are treated in a sustainable manner. While a number of these might arguably fall into the category of global commons, as suggested by the Convention on Biological Diversity (CBD) (a point also contested within the text of the convention), none is as central to the political economy of many countries as production of timber and conversion of land. Moreover, while sovereignty considerations do enter into access to genetic resources, with nominal limits to access addressed in the CBD, neither considerations of sovereignty nor global commons appears relevant to any of the other secondary benefits provided by forests. For the time being, these might be thought of as positive externalities for which no one pays but from which everyone benefits. In political terms, concentrated interests and the protection of national control far outweigh the diffuse and scattered interests that the world may have in these secondary benefits.

As defined in the 1993 Helsinki Declaration of the Ministerial Conference on the Protection of Forests in Europe, sustainable forest management is: "the stewardship and use of forests and forest lands in a way, and at a rate, that maintains their biodiversity, productivity, regeneration capacity, vitality and their potential to fulfil, now and in the future, relevant ecological, economic and social functions, at local, national and global levels, and that does not cause damage to other ecosystems" (*International Trade Forum*, 2002). Negotiations over an international forest convention, which would establish some level of harmonized SFM standards among countries, failed repeatedly during the 1990s. The "Non-legally binding authoritative statement of principles for a global consensus on the management, conservation and sustainable development of all types of forests" (UNCED, 1992), signed at the 1992 Earth Summit in Rio de Janeiro, contained no provisions for an international law to regulate forestry. At the time, national governments were leery of being bound to a single set of rules, while many environmental nongovernmental organizations (NGOs) believed that an agreement would only foster increased international trade in timber and even higher rates of deforestation than had been taking place.

Unwilling to lose the momentum generated by the UNCED Forest Principles, however, in 1993 an Intergovernmental Working Group on Global Forests (IWGF) was created (the word *global* was later dropped). A joint initiative of the Canadian and Malaysian governments, the IWGF held a series of meetings of experts and officials from fifteen key forest countries and several NGOs to facilitate dialogue and consolidation of approaches to the management, conservation and sustainable development of the world's forests. By the second meeting, attendance had expanded to include technical and policy experts from thirty-two countries including Brazil, the United States, Indonesia, Finland, Sweden, the Russian Federation, Japan, Gabon, five intergovernmental organizations and eleven NGOs (International Institute for Sustainable Development [IISD], n.d. a). At the end of 1994, the final report of the IWGF was presented to the UN Commission on Sustainable Development (CSD) which, at its third meeting in 1995, proposed to establish an ad hoc Intergovernmental Panel on Forests (IPF) to further examine issues and develop proposals and recommendations. The IPF held four subsequent meetings through 1997, when its final report was submitted to the CSD (IISD, n.d. b). As a followup to the work of the IPF, in 1997, the UN Economic and Social Council (ECOSOC) established the Intergovernmental Forum on Forests (IFF), which pursued the work of the IPF and developed additional action proposals. Ultimately, the IPF and IFF together issued 270 proposals for action (UNFF, n.d. a). And, finally, in 2000, ECOSOC established a permanent entity, the UN Forum on Forests (UNFF), to build on the work of its predecessors (UNFF, n.d. b). None of these initiatives led, however, to a Global Forestry Agreement, and therein lies a tale.

Initially, the United States was a strong supporter of such an agreement, in the view that tropical deforestation represented a major contributor to global warming. Preferring to see other countries, especially less developed ones (LDCs), reduce their emissions, the UNCED Forest Principles were the most to which the LDCs would agree. After UNCED, a number of governments, including European, LDC, and Canadian supported a global agreement but, by 1996, the U.S. position had changed completely, as boreal and temperate forests were included in the remit of the various panels and forums addressing deforestation, and industry opposition grew. Environmental organizations, too, were opposed to a global forest agreement and wished, instead, to see forest conservation addressed through the CBD (Fogel, 2002, p. 129).

The nail in the coffin, as it were, occurred when the Kyoto Protocol became the locus of global forestry regulation, under the rubric of LULUCF, or Land Use, Land Use Change, and Forestry. In effect, the United States and several other countries began to see in forests the possibility of sequestering carbon and avoiding the need to reduce greenhouse gas emissions in other sectors, such as transport and industry. Cathleen Fogel (2002) has nicely documented the logic behind this shift from conservation of standing forests to sequestration through replanting forests already cut down. Through the Clean Development Mechanism (CDM) and other modalities, carbon emissions in the form of standing trees will be traded, and sustainable forestry will become something quite different from what was originally envisioned. While a few countries, such as Canada, continue to call for an agreement in order to override the proliferation of certification schemes, for the moment, global public forestry regulation appears quite unlikely.

The result of almost fifteen years of international deadlock on SFM has been a proliferation of private and national regulatory projects for sustainable forestry—as many as fifty by some counts—and the stakes are higher than ever. Forests have, historically, been subject to considerable state regulation, if only because, until the mid-nineteenth century, timber played a major role in military affairs. Forests were often the property of kings and aristocrats, who were zealous about protecting them, and governments regarded forests as integral to projects of national development. And forests occupy national territory and are regarded as sovereign resources. There is considerable competition among the various forest regulation codes. The one that is most widely adopted and accepted, by both consumers and producers, will acquire a monopoly position in the market for such regulation and become the basis for international forestry law.

As has been the case in a number of other environmental sectors, privatized efforts to regulate forestry practices have come to rest largely on the tools of trade. For better or worse, however, both international trade law and the advocates of free trade stand in opposition to such international regulation. Public international forestry law would mandate some degree of harmonization of forestry practices yet, just as in the case of labor law, free trade advocates generally argue that this would amount to a form of cultural imperialism. They are, therefore, opposed to the inclusion of environmental regulations in trade agreements (Bhagwati, 1993, 2002). In the absence of such harmonization, individual states find themselves in a weak position to impose their own municipal standards on forestry imports in an effort to encourage more sustainable practices in the country of origin, for two reasons. They might then be found in violation of WTO rules that forbid process standards as nontariff barriers to trade (see, e.g., Mayer & Hoch, 1993). Countries with lower regulatory standards might also be able offer timber at lower costs. Countries can impose their own domestic standards but these are likely to increase variable costs; paradoxically timber producers in high-cost countries such as Canada are demanding international harmonization so that they may remain competitive (Barron, 1997). These reasons are why the agreement presented at Rio was characterized as Forestry Principles, rather than as a binding convention; as principles, countries could choose to practice them or not. Most have chosen not to.

The resulting lacuna has led both activists and the timber industry to search for other means of regulating forest practices at the global level. Timber company brands are hardly as ubiquitous as those of clothing manufacturers, for example, so that consumer awareness is a less powerful tool. The global market structure of the timber trade is much more fragmented, too. Activists have chosen, therefore, to pursue a double-pronged strategy. First, they are putting pressure on retailers and do-it-yourself stores in Europe and North America to sell only sustainably managed and produced lumber and, of course, to inform consumers that they are doing so. Demand from these retailers, it is hoped, will induce wholesalers and producers to seek sustainable timber for sale to contractors and do-it-yourselfers. But many timber companies and governments are reluctant to hop on activist bandwagons, regarding their standards as being too high or potentially difficult to control. Consequently, the business equivalents of corporate social responsibility are also on offer. These are regulatory standards formulated by timber associations or companies themselves. Activists dislike such standards because they do not trust corporations to police themselves. Table 3.1 lists a number of public and private programs offering forestry regulation. These efforts and projects to regulate forestry fall into several different categories, running along one spectrum from public to private, and a second from political to economic. The resulting four categories are displayed in Table 3.2.

Many of the projects listed in Table 3.1 seek to regulate economic activities through certification. There are three types of market-based certification. First party labeling, the most common and simplest approach, entails producer claims about a product, such as recyclable, ozone-friendly, nontoxic or biodegradable. In the absence of a mechanism for verifying these claims, the only guarantee that the product performs accordingly is the producer's reputation. Second party labeling is conducted by industry-related entities, such as trade associations, which establish guidelines or criteria for making claims about the product. Once the standards are met or the guidelines followed, an industry-approved label is placed on the product stating or verifying the product's environmentally friendly qualities. In this instance, corporate members of the certifying organization will seek to ensure the label's value, and to mandate its use, so that no single producer will have an advantage over any other. Third party, or independent, labeling is performed by either a governmental agency, a nonprofit group, a for-profit company, or an organization representing some combination of these three. As with second party type, third party labeling programs set guidelines that products must meet in order to use their label. They may also conduct audits in order to ensure compliance with the guidelines. As the name implies, third party organizations are not affiliated with the products they label (Caldwell, 1998; Bass & Simula, 1999).

Timber certification comes in two forms. Forest management certification involves assessment of forestry practices by a company, community, or other organization according to a set of predetermined standards. The focus of such certification may be an individual forest or a set of forests managed by an organization. It may also be conducted regionally or nationally, depending on the management structure of the forestry and timber sectors. Wood product certification involves an inspection of the chain of custody to follow wood throughout the commodity chain. This is done by auditing individual organizations at each step of the chain to determine whether or not they are using materials from certified sources (Oliver, 1996).

Table 3.1. Current initiatives in sustainable forestry regulation

	NGOs	States	Corp.	
Kyoto Protocol		✔		Establish terms and conditions to meet provisions of Kyoto Protocol regarding management of forests and their role as carbon sinks
Intergovernmental Working Group on Global Forests (1993–1994)		✔		Created to develop a scientifically-based framework of criteria and indicators for the conservation, management, and sustainable development of boreal and temperate forests.
UN Intergovernmental Panel on Forests (IPF) (1995–1997)		✔		Created by the UN Commission on Sustainable Development as an open-ended ad hoc group to pursue consensus and coordinate proposals to support the management, conservation, and sustainable development of forests.
Intergovernmental Forum on Forests (IFF) (1997–2000)		✔		Follow-up to the IPF created by ECOSOC to pursue further proposals for action to governments, international organizations, private sector entities, and all other major groups on how to further develop, implement, and coordinate national and international policies on sustainable forest management.
UN Forum on Forests (UNFF) (2000–present) (www.un.org/esa/forests)		✔		Created as the permanent intergovernmental body responsible for overseeing the implementation of the IPF/IFF Proposals for Action and enhancing cooperation and international forest policy dialogue.
International Tropical Timber Organization (1985–present) (www.itto.or.jp)	observers	✔	observers	Created in 1985 to provide international reference document upon which more detailed national standards could be developed to guide sustainable management of natural tropical forests.
Center for International Forestry Research (CIFOR)	✔	✔		Established to improve the scientific basis for ensuring the balanced management of forests and forest lands; develop policies and technologies for sustainable use and management of forest goods and services.

continued

Table 3.1. *Continued*

	NGOs	States	Corp.	
International Organisation for Standardization ISO 14001 (www.iso.ch)		✔	✔	ISO series provides a framework for an organization to use to identify and address the significant environmental aspects and related impacts of its activities, products, and services
World Commission on Forests and Sustainable Development (1996–1999)	✔	✔	✔	Independent commission which held hearings to achieve policy reforms aimed at reconciling economic and environmental objectives for sustainable management of global forests
Rainforest Action Network (www.ran.org)	✔			Old Growth Campaign promotes consumer boycotts of companies that log and sell products from old growth forests
Smart Wood (1989–present) (www.smartwood.org)	✔			Established by the Rainforest Alliance to provide certification to all types of operations in all types of forests. FSC accredited.
Forest Stewardship Council (FSC) (www.fscoax.org)	✔		✔	Created in 1993 to establish internationally-recognized principles and criteria of forest management as a basis for accrediting regional certifiers
Scientific Certification Systems (Oakland, CA) (www.scs1.com/forestry.shtml)			Private firm	Forest Conservation Program evaluates forest management against objective and regionally appropriate principles of sustainable forestry; FSC certified
SGS Qualifor (Oxford, UK) (www.qualifor.com)			Private firm	Carbon Offset Verification Service assesses, surveys, monitors, and certifies project development and management
Pan-European Forest Certification (1999–present) (www.pefc.org)			✔	Created to provide certification of forests according to the Pan-European Criteria as defined by the resolutions of the Helsinki and Lisbon Ministerial Conferences of 1993 and 1998 on the Protection of Forests in Europe

continued

Table 3.1. *Continued*

	NGOs	States	Corp.	
Sustainable Forestry Initiative (1995–present) (www.aboutsfi.org)			✔	Established by American Forest and Paper Association to provide standard of environmental principles, objectives and performance measures that integrates growing and harvesting of trees with the protection of wildlife, plants, soil and water quality, and other conservation goals for international application
African Timber Organisation		✔		Pan-African timber trade organization with 13 member countries developing standards for sustainable forest management that could form eventual basis for certification program
Malaysian National Timber Certification Council (www.mtcc.com.my)		✔		Quango established to administer voluntary third-party certification of forests in Malaysia. Cooperates with FSC
Lembaga Ekolabel Indonesia (1998–present) (www.lei.or.id)	✔			Certifying organization for Indonesian forests, works in cooperation with FSC
International Forest Industry Roundtable		✔		Proposal for an international mutual recognition framework for national forest forest certification programs is in the works
The BMZ/GTZ Forest Certification Project (www.gtz.de/forest_certification)		✔		German government-owned corporation which provides training and support for information, capacity building, participation and networking for better communication and cooperation of those involved in certification processes
Initiative zur Föderung nachhaltiger Waldbewirtschaftung (IFW)			✔	Dual process of certification whereby nationally accredited bodies within timber exporting nations would certify that producers have met high standards standards of forest management, for European label

Sources: Evans, 1996; CIFOR, 2004; SGS Quzli for 2004; IISD, 2003; UNFF, 2004; other forestry web sites.

Table 3.2. Institutional form of sustainable forestry regulation

	Political	Economic
Public	**Interstate** UNCEF Forestry Principles	**Activist** Forest Stewardship Council
Private	**Transnational** International Forestry Industry Roundtable	**Private** ISO 14001

Finally, the entity responsible for certification may be either independent (third party) or national. In the former case, standards are usually formulated by an organization, whether public, private, or nonprofit, with no ties to the companies whose practices and products are subject to certification. The standard-setting organization then authorizes other independent entities to act as certifiers. Alternatively, certification standards may be devised by national or international associations whose members are owners of forests and producers or sellers of wood products. In the latter case, the association may certify companies, or the task may be passed on to some kind of independent certification body. In all cases, the company or individual seeking certification for a property pays the independent auditor to examine, assess, and certify the forest. Once approved, certified timber companies, producers, and products are permitted to display an eco-label intended to inform consumers that SFM standards have been met (Oliver, 1996). Clearly, however, the credibility of the eco-label is no easy thing for a consumer to assess.

Activist certification

The first forestry certification programs opened for business in 1989. In response to a 1988 study of the process by the International Tropical Timber Organization, the Rainforest Action Network (RAN) initiated successful U.S. consumer campaigns to boycott the import and use of all tropical timber except that produced from sustainably managed forests. In 1989, the Rainforest Alliance launched Smart Wood, the first industry-independent certification program. At the same time, the Rogue Institute in Ashland, Oregon began a verification program to promote environmentally sensitive timber production as an alternative to clearcut logging in the southern part of the state. Other groups concerned with sustainable forestry included the Sierra Club, Friends of the Earth, Greenpeace, the National Wildlife Federation, and the Woodworkers Alliance for Rainforest Protection (WARP), the last representing concerned wood users, as well as several smaller grassroots forests groups, indigenous peoples, social organizations, timber producers, and timber retailers from several countries.

The Forest Stewardship Council (FSC) is the best known of the private nonprofit certification groups. The FSC was launched in 1993 in Washington, DC by environmental groups, the timber industry, foresters, indigenous peoples and community groups from twenty-five countries, with initial funding provided primarily by the Worldwide Fund for Nature/World Wildlife Fund (WWF). An interim board was elected, a mission statement adopted, and draft Principles and Criteria for Forest Management formulated soon thereafter. The FSC was originally based in Oaxaca, Mexico, but has recently moved its central office to Bonn, Germany, where it will be better positioned to compete with other standard-setting organizations. It is a membership organization comprised of three equally weighted chambers—environmental, social, and economic—and membership within each chamber is also equally weighted between North and South. As the FSC's web site puts it:

- The Environmental Chamber includes nonprofit, nongovernmental organizations, as well as research, academic, technical institutions and individuals that have an active interest in environmentally viable forest stewardship;
- The Social Chamber includes nonprofit, nongovernmental organizations, as well as research, academic, technical institutions and individuals that have a demonstrated commitment to socially beneficial forestry.
- The Economic Chamber includes organizations and individuals with a commercial interest. Examples are employees, certification bodies, industry and trade associations (whether profit or nonprofit) wholesalers, retailers, traders, consumer associations, and consulting companies. (FSC, 2002a)

Each chamber represents 33 percent of the vote at annual meetings, and the board of directors has rotating members reflecting these interests. By 2001, the FSC was an internationally recognized organization with 448 members in 56 countries, 221 in the economic chamber, 86 in the social chamber, and 174 in the environmental chamber (Meridian Institute, 2001, p. 20).

With international governmental processes in apparent stalemate, the FSC has been seen by many as the magic bullet, a market driven mechanism able to fill a critical niche towards achieving sustainable forest management where governments cannot. According to its mission statement:

> 1. The Forest Stewardship Council A.C. (FSC) shall promote environmentally appropriate, socially beneficial, and economically viable management of the world's forests.
>
> 2. Environmentally appropriate forest management ensures that the harvest of timber and non-timber products maintains the forest's biodiversity, productivity, and ecological processes.
>
> 3. Socially beneficial forest management helps both local people and society at large to long term benefits and also provides strong incentives to local people to sustain the forest resources and adhere to long-term management plans.
>
> 4. Economically viable forest management means that forest operations are structured and managed so as to be sufficiently profitable, without generating financial profit at the expense of the forest resource, the ecosystem, or affected communities. The tension between the need to generate adequate financial returns and the principles of responsible forest operations can be reduced through efforts to market forest products for their best value. (FSC, 2002b)

The FSC has developed and adopted global Principles and Criteria for Forest Management and it accredits certifying organizations that agree to abide by these Principles and Criteria. Purportedly, the FSC also monitors the operations and portfolios of such certifying groups on an annual basis. In cooperation with lumber retailers, the FSC creates Buyers Groups in consuming countries. Members of these groups are committed to selling only verified sustainably produced timber in their stores (FSC, 2002). As of January 2003, the FSC had granted 466 forest management certificates in 56 countries, covering almost 77 million actors, and 2,801 chain of custody certificates in 67 countries (FSCUS, 2003).

The actual ecological and social outcomes triggered by the FSC system are not yet clear, however, and have not yet been well studied (but see Mater, Price, & Sample, 2002). Some indications are that in some locations, the system is not leading to ecological or social outcomes that exceed those already required by existing governmental policies. In other instances, FSC standards may not actually be implemented by producers, due to the weak institutional base of the FSC. Funding and personnel to monitor implementation are scarce, and penalties for failing to observe the rules are few (e.g., Freris & Laschefski, 2001). Moreover, the large financial stakes involved have led forest products companies to become actively involved in standard setting and implementation activities in several countries such as Sweden and Canada. This appears to be leading to a consensus rather than science-based approach to standard setting in order to make the standards achievable, and thus to ensure that the large and growing market demand will indeed be met.

An additional challenge to the FSC's success may be the broader trend toward green labeling that it has inspired. Its forest product certification program has triggered numerous corporate and government responses, and considerable alarm. The large financial stakes involved have led forest products companies to become actively involved in standard setting and implementation activities in a number of other countries such as Sweden, Indonesia and Malaysia. As is evident in Table 3.1, a growing number of organizations including the American Forest Products Association, and the Canadian Pulp and Paper Association, in conjunction with the International Organization for Standardization, have developed certification programs (e.g., CSFCC, 2003; Lipschutz, 2001). In particular, these industrial projects may reflect an attempt to expropriate forest product certification processes, principles and discourse from the FSC and other environmental organizations (Hauselmann, 1997).

Industry certification

The International Organization for Standarization (ISO), based in Geneva, is a quasi-governmental body with member organizations in 119 countries. It is the official standard-setting and labeling body recognized by the WTO and other international agencies. Founded in 1946, "ISO's mission is to promote standardisation and related activities in the world with a view to facilitating the international exchange of goods and services and to developing cooperation in the spheres of intellectual, scientific, technological and economic activity by developing worldwide technical agreements which are published as international standards" (Hauselmann, 1997, p. 3). With an annual operating budget of $125 million provided by governments and corporate members, the ISO is far larger than the FSC and other comparable certifying organizations. Around the world, it hosts as many as ten standards-setting meetings each day (Hauselmann, 1997, p. 3). Unlike the FSC, the ISO is frequently the recipient of praise and support by governments and most of the forest products industry.

The organization only provides the context within which standards can be negotiated and promulgated; it does not engage in policing corporate behavior, enforcing standards, or penalizing violators. In fact, individual corporations generally devise their own internal performance programs which are vetted and certified by an authorized company or organization. In other words, a producer whose program receives second-party certification from an ISO-approved auditor is, for the most part, self-regulating and responsible for seeing that it meets the terms of its programs.

Historically, the ISO has neither worked on nor developed competency in either environmental or forestry issues. Until the early 1980s, it limited itself to purely technical standards, such as the size of nuts and bolts (Hauselmann, 1997). The demand for environmental standards grew out of a concern that these might be imposed "from above" as a result of interstate agreements and conventions. Growing public agitation over the absence of any environmental considerations in the General Agreement on Tariffs and Trade (GATT) and, later, the WTO also contributed to the ISO's entry into the environmental standards business (Lally, 1998, p. 4).

In 1993, the ISO initiated a process of developing a new ISO 14000 Series of Environmental Management Systems standards. This was intended to build on the success of the ISO 9000 Quality Management Systems, which are de facto requirements for companies engaging in most sectors of international trade (Cascio, Woodside, & Mitchell, 1996). Those standards are driven by the market and based entirely on self-regulation (Lally, 1998, p. 3). The ISO adheres to the Environmental Management System approach. This approach differs from the FSC's Principles, Criteria and Standards for forest management in that Environmental Management Systems only prescribe internal management systems for companies that wish to continuously improve upon an environmental performance level which they themselves define. Adherence to externally agreed standards (ostensibly set by all interested stakeholders) is not required (as it is in the FSC). Furthermore, the ISO has no adequate mechanism to either ensure corporations' compliance with or the effectiveness of their individual action plans, or to control the use (or misuse) of logos and certification marks. In other words, ISO 14000 involves only *first party* certification.

As a result, there is, according to one observer, a "potential for confusion . . . this situation is worse in the case of forest management certification, where some economic interests are seeking to use the ISO framework to develop a forestry-specific application of the Environmental Management System (EMS) approach in order to counter an existing and operational environmental labeling scheme—that of the Forest Stewardship Council" (Hauselmann, 1997). Although the ISO has well-developed procedures on consensus and participation, these have not been well followed in creating the ISO 14000 Series. Environmental organizations have not been allowed to attend standards-setting meetings

(Hauselmann, 1997), ostensibly to avoid politics. Instead, corporate forest product industry efforts seem to be aimed at imbuing the ISO with an aura of scientific, technical and social legitimacy, all the while maintaining a near perfect level of control.

Nevertheless, forest industry members and supporters of the ISO 14000 series are using the discourse developed by the FSC and environmental groups to describe their systems approach in terms uncannily similar to those adopted by the FSC. For example, a 1997 press release issued by the Canadian Sustainable Forestry Certification Coalition (an industry group), promoting ISO forest certification, claims that: "we have identified the background information that forestry organizations will find useful as they implement and progressively improve upon their environmental management system. This major step forward in relating the key elements of the ISO standard in the context of a range of international forest management measures will further the UN Agenda 21 goal of promoting sustainable development" (CSFCC, 1997).

Some ISO members continue as well to actively push forward the development of international ISO forest management system standards. Some are concerned that certification might obstruct free trade and are active at the WTO Environment Committee to limit the definition and mutual recognition of eco-labels by GATT country signatories. Consequently, although timber products may carry ISO certification, what lies behind the label is none too clear.

Transnational harmonization of national standards

The large number of forestry certification programs has been particularly frustrating to national timber industry associations, who see fragmented privatization as disadvantageous to their members. Something of a backlash has developed among the national associations, in particular, who would prefer to retain their own national certifications systems but have them recognized by other national associations. Because the likelihood of formulating a global forest convention, much less ratifying one, is so low, the industry strategy has been to seek mutual recognition of competing standards. As the Canadian Sustainable Forestry Certification Coalition, composed of national, provincial, and sectoral associations, has put the case for mutual recognition: "Although nice in concept, it is unlikely that one standard could ever speak to the diversity of forest types and ecosystems across North America, to the diversity of tenure systems, to public ownership, to private ownership, to the different needs and operating systems within a business, including their varied sources of wood supply, or to the different needs and priorities of the users of wood products. While one standard could run the risk of not speaking to the forest management realities of many operations, many standards will likely result in more widespread application, and in the end, more improvements in forest management" (CSFCC, 2002).

One transnational harmonization scheme is the International Forest Industry Roundtable's (IFIR) mutual recognition project. IFIR is a self-described independent network of industry associations, with members from Argentina, Australia, Brazil, Canada, Chile, Finland, France, Malaysia, Mexico, Norway, New Zealand, South Africa, Sweden, the United Kingdom, and the United States. In 1999, it established a working group to develop an International Mutual Recognition Framework for national forestry certification standards. The objective of this initiative is to "provide a critical mass of credibly certified wood products by recognising that *different* certification systems can provide *substantively equivalent* standards of sustainable forest management. Mutual recognition would set a high threshold for entry for participating standards, while enabling the use of standards that accommodate local and regional circumstances. By providing a process to *differentiate credible from non-credible* certification standards, mutual recognition would use market forces to provide a range of certification standards that will assure customers that their wood product purchases contribute to sustainable forest management" (Griffiths, 2001, p. 3; emphasis in original).

Although it is not stated outright, mutual recognition of national standards may be aimed against the Forest Stewardship Council, which is beginning to look like a default global standard setter, if only because of its broad membership and environmentalist credentials (Griffiths, 2001, p. 8). There is also fear of the "potential *imposition* of 'mandatory' solutions via government regulation at the national or international level" (Griffiths, 2001, p. 8) if the industry is unable to self-regulate. As of this writing, the members of IFIR continue to negotiate over the matter.

Can private forestry certification work?

Does private certification of SFM provide an adequate substitute for public regulation? For the most part, the jury is still out on this question. In his introduction to this volume, Michael Hatch proposed five criteria for assessing the relative usefulness of policy instruments for environmental protection. Here, we evaluate the forestry certification schemes described in this chapter in terms of those criteria.

1. Environmental effectiveness: According to IFIR, global sales in the forest products business amount to about $500 billion per year, of which some 30 percent enters international trade (Grifflths, 2001, p. 5). The market for certified timber is, as yet, only a small fraction of this. Estimates of the total area of certified forests worldwide range from 265 to almost 500 million acres (about 2–5 percent of the world's forests; FAO, 2001, p. xii; CSFCC, 2002). It is much more difficult, however, to find hard data on the impacts and effectiveness of such certification on the health of those forests that have been certified by the various programs. The vast majority of certified forests are in industrialized countries, and it appears that most of those forests were already being managed close to certifier standards. Furthermore, the long-term consequences of certification, especially for natural forests (whether old-growth or new-growth) cannot be assessed until a significant fraction of a harvesting cycle has passed. Consequently, for the time being there appears to be no way to determine whether certification, as a policy instrument, offers a viable long-term means of protecting the environment (Bass, et al., 2001).

2. Economic efficiency: The economic efficiency of private forestry certification rests on answers to three questions. First, does it reduce the costs of regulating forestry practices, thereby motivating both governments and timber companies to subscribe to and support certification? Second, what is its impact on forestry costs themselves, that is, how much does certification add to a timber company's costs of doing business? And, third, does certification result in a benefit to the consumer and the global public? None of these questions are either straightforward or easy to answer. In principle, competition among the various programs ought to result in a lowering of certification costs, as well as elimination of deadweight costs associated with state-based regulation. In practice, however, there has not as yet emerged a single world standard for SFM, although there are some indications that the FSC's criteria are becoming the de facto standard. But the different standard-setting organizations also have different, and sometimes conflicting, objectives. Some timber companies may find it necessary to apply for multiple certifications, and that certainly will not be economically efficient compared to a global standard.

Clearly, there are costs of meeting certification standards for SFM. The growing demand for certified lumber and wood products has outstripped supply, and this has made it possible to sell certified goods at a premium. As certification becomes more widespread, however, this premium will decline and, at the margin, will provide little or no benefit to the producers in the timber commodity chain. At that point, all else being equal, the benefits of sustainable forestry will have been internalized and socialized, with the global public and environment as the beneficiaries. But, if sustainable forestry is voluntary and coverage does not extend to all forests, whether North or South, there will be strong incentives by noncertified producers to free-ride on the global trade system. Recall, moreover, that the WTO forbids discrimination against substantially equivalent products on the basis of production method.

Finally, certification reduces transaction costs for the consumer of lumber and wood products, although that savings might be wiped out by the premium associated with certification. The global benefits of sustainable forestry will be imperceptible to the individual consumer and, once again, the temptation to free ride may be great. It is one thing to tack a 10 percent green surcharge on a piece of furniture that may cost between $100 and $1,000; it is quite another to charge an extra 10 percent on a $20,000 remodeling job or a $300,000 house, which may make the difference between obtaining a mortgage and having a loan application turned down.

3. Political efficiency: For the moment, there is no international convention or agreement stipulating standards for sustainable forestry that could be carried into national regulations and policies. Although there are a few governments who would prefer to have such regulations in place, timber producers and associations have made clear their preference for self-regulation. Under ideal conditions, this might be a second-best solution to the problem, but the proliferation of certification schemes and, especially, fragmentation amongst national standards, suggests that we are far from even this. It would be unreasonable, of course, to expect completely effective certification and sustainable forestry even under a global convention, but there would, at least, be a single framework from which to develop the required policy tools.

4. Administrative efficacy: It is difficult to imagine any timber certification system being administratively efficacious. The challenge here is to find some sort of basis on which to audit, assess, and monitor practices taking place in a range of different forests

(old-growth, second-growth, new-growth, etc.) managed in many different ways (community, private, public, etc.) under contrasting environmental conditions (tropical, temperate, boreal, etc.). Given that timber from these many different sources enters into international trade, how are judgments of essential equivalence to be made? The administrative task would be made simpler under a single standard setter, whether that be the FSC or the FAO or even IFIR.

5. Technological innovation: If there are any innovations driving the movement for sustainable forestry, they are social. To be more precise, the movement is driven by two motivations: habitat maintenance, on the one hand, and consumer consciousness, on the other. Protection of forests and habitat could be accomplished by any number of strategies, many of which have been tried and a number of which have failed. Because the market is such a powerful force in environmental degradation, and efforts to exclude the market from environmentally-sensitive areas have been, in many cases, such failures, the temptation to harness the market in the service of environmental protection seems both innovative and promising. The consumer appears to be the lever that can move industry toward social responsibility. By appealing to the interests of both—the consumer's in environmental protection and the corporation's in increased profits—certification looks like the magic formula that will make it possible to do well by doing good.

There is, however, an unrecognized trap hiding in the market-based approach to regulation of sustainable forestry, in particular, and of social and environmental externalities, in general: Markets are particularly weak arenas in which to seek political goals. While some argue that there is such a thing as private political power, which can be accumulated through the market, this seems a somewhat oxymoronic concept. Politics is, by definition, a public, collective endeavor, while markets involve private exchange between individuals. Politics is based on the visible aggregation of the power, which markets eschew. Politics through market-based methods, which is what private certification amounts to, rests primarily on attempts to alter the preferences of large numbers of consumers in order to put pressure on producers. Because consumer preferences are not political and are strongly influenced, if not determined, by the very system of production and consumption that motivates the social disruption and externalities of concern, there is a certain tautological process at work here. If capital is able to acquire political power, it is more a form of displacement than an alternative: the corporate citizen becomes, in a sense, a franchisee able to cast a vote using its dollars.

The relevant question here, then, is not about best or second-best solutions to what appears to be a largely technical problem about maintaining best practices in forestry or any other environmental sector. If sustainable forestry is as critical to the sustenance of biological diversity and the environment as is often claimed, institutions that emerge from and through the market are unlikely to provide the necessary normative and legal structures that will last through much more than a few of the required harvesting cycles. State bureaucracies and international organizations do not, for the most part, have much greater staying power, although there are a few that have been in existence for centuries. Indeed, if and when the alternative policy instruments for environmental protection are put into place, the task of protecting the environment will have only just begun.

Note

*This chapter is based on material drawn from Ronnie D. Lipschutz, *Regulation for the Rest of Us? Globalization, Governmentality, and Global Politics* (forthcoming). Support for this research has been provided by Sokka University America, the Aspen Institute, the University of California Institute for Labor and Employment, the University of California Institute on Global Conflict and Cooperation, and the Academic Senate and Social Sciences Division of University of California-Santa Cruz.

References

Barron, D. (1997, May). CPPA calls for a legally-binding forest convention. *On Paper.* Canadian Pulp and Paper Association. Retrieved February 6, 2003, from http://www.cppa.org/english/cppa/onpaper/9705/.

Bass, S., & Simula, M. (1999). Independent Certification/Verification of Forest Management. Background paper prepares for the World Bank/WWF Alliance Workshop, Washington, DC, November 9–10, 1999. Retrieved February 6, 2003, from http://www.gtz.de/forest_certification/download/d28.pdf.

Bass, S., M. Grieg-Grzn, M. Markopoulos, S. Roberts, and K. Thornber. (2001, May). *Certification's impacts on forests, stakeholders and supply chains.* Stevenage, Herts.: Earthprint. Retrieved Februay 5, 2003, from http://www.iied.org/psf/publications_def.html#cert.

Bhagwati, J. (1993). Trade and the environment: The false conflict? In D. Zaelke, P. Orbuch, & R. F. Houseman (Eds.), *Trade and the environment—Law, economics, and policy.* Washington, DC: Island Press.

Bhagwati, J. (2002). *Free trade today.* Princeton, NJ: Princeton University Press.

Braithwaite, J., & Drahos, P. (2000). *Global business regulation.* Cambridge, UK: Cambridge University Press.

Bromley, D. W. (Ed.). (1992). *Making the commons work.* San Francisco: ICS Press.

Caldwell, D. J. (1998, October 30). Ecolabeling and the regulatory framework—A survey of domestic and international fora. Prepared for the Consumer's Choice Council, Washington, DC (discussion draft). Retrieved February 6, 2003, from http://www.consumerscouncil.org/policy/ecolab1.htm.

Cascio, J., Woodside, G., & Mitchell, P. (1996). *ISO 14000 guide: The new international environmental management standards.* New York: McGraw Hill.

Canadian Sustainable Forestry Certification Coalition (CSFCC). (2003). Certification Status and Intentions in Canada. Retrieved June 8, 2003 from http://www.sfms.com/status.htm.

Canadian Sustainable Forestry Certification Coalition (CSFCC). (2002). Mutual recognition. Retrieved February 2, 2003, from http://www.sfms.com/recognition.htm.

Canadian Sustainable Forestry Certification Coalition (CSFCC). (1997, November). ISO Forestry Working Group completes technical report. ISO/TC207/WG2, *Forestry.* Retrieved September 9, 1999, from http://www.sfms.com/rece7l.htm.

Center for International Forestry Research (CIFOR). (2004). CIFOR Home Page. Retrieved June 8, 2004, from http://www.cifor.cgiar.org/.

Cutler, A. C. (1999). Private authority in international relations: The case of maritime transport. In A. C. Cutler, V. Haufler, & T. Porters (Eds.), *Private authority and international affairs* (pp. 283–329). Albany, NY: State University of New York Press.

Evans, Bryan. (1996). "Technical and Scientific Elements of Forest Management Certification Programs," Paper prepared for the conference on Economic, Social and Political Issues in Certification of Forest Management, Universiti of Pertanian, Malaysis, May 12–16. Retrieved June 8, 2004, from http://www.forestry.ubc.ca/concert/evans.html.

Fogel, C. (2002). *Greening the earth with trees: Science, Storylines and the construction of international climate change institutions.* Unpublished doctoral dissertation, University of California, Santa Cruz.

Food and Agriculture Organisation. (2001). *State of the world's forests 2001,* Rome. Retrieved February 7, 2003, from ftp://ftp.fao.org/docrep/fao/003/y0900e/y0900e00.pdf.

Forest Stewardship Council. (2002a). The economic, social, and environmental chambers. Retrieved February 7, 2003, from http://www.fscoax.org/principal.htm.

Forest Stewardship Council. (2002b). Forest Stewardship Council A.C. by-laws. Document List, revised November. Retrieved February 7, 2003, from http://www.fscoax.org/principal.htm.

Forest Stewardship Council United States. (2003, January 15). Results/Impacts. Retrieved February 7, 2003, from http://fscus.org/results_impact/index.html.

Freris, N., & Laschefski K. (2001). Seeing the wood from the trees. *The Ecologist, 31*(6). Retrieved June 8, 2004, from http://www.theecologist.org/archive_article.html?article=106.

Griffiths, J. (Ed.). (2001). Proposing an international mutual recognition framework. Report of the Working Group on mutual recognition between credible sustainable forest management certification systems and standards, International Forest Industry Roundtable. Retrieved February 4, 2003, from http://www.sfms.com/pdfs/ifirframework.pdf.

Haas, P. (Ed.). (1992). Knowledge, power, and international policy coordination. *International Organization, 46*(1), special issue.

Hauselmann, P. (1997). *ISO inside out: ISO and environmental management.* Surrey, UK: WWF, International Discussion Paper.

International Institute for Sustainable Development (IISD). (2003). "A Brief Introduction to Global Forest Policy." Retrieved June 8, 2004, from http://www.iisd.ca/process/forest_desertification_land-forestintro.htm.

International Institute for Sustainable Development (n.d. a.). Intergovernmental Working Group on Forests. Retrieved February 7, 2003, from http://www.iisd.ca/linkages/forestry/iwgf.html.

International Institute for Sustainable Development (n.d. b.). A brief history of the Intergovernmental Panel on Forests. Retrieved February 7, 2003, from www.iisd.ca/linkages/forestry/ipfhist.html.

International Trade Forum. (2002). "Certification Concepts Defined," Issue 2. Retrieved September 4, 2003, from http://www.tradeforum.org/news/fullstory.php/aid/281.html.

Lally, A. P. (1998). ISO 14000 and environmental cost accounting: The gateway to the global market. *Law and Policy in International Business, 29*(4), 501–538.

Lipschutz, R. D. (2000). *After authority—War, peace and global politics in the 21st century.* Albany, NY: State University of New York Press.

Lipschutz, R. D. (2001). Why is there no international forestry law? An examination of international forestry regulation, both public and private. *UCLA Journal of Environmental Law & Policy, 19*(1), 155–182.

Lipschutz, R. D. (2002). Doing well by doing good? Transnational regulatory campaigns, social activism, and impacts on state sovereignty. In J. Montgomery & N. Glazer (Eds.), *Sovereignty under challenge: How governments respond* (pp. 291–320). New Brunswick, NJ: Transaction.

Lipschutz, R. D. (forthcoming). Civil society, globalization, and the environment. In K. Crowley (Ed.), *Globalization and the environment.* New York: Nova Sciences.

Lipschutz, R. D., with Mayer, J. (1996). *Global civil society and global environmental governance.* Albany, NY: State University of New York Press.

Mater, C. M., Price, W., & Sample, V. A. (2002, June). Certification assessments on public and university lands: A field-based comparative evaluation of the Forest Stewardship Council (FSC) and the Sustainable Forestry Initiative (SFI) Programs. Pinchot Institute for Conservation, Washington, DC. Retrieved February 7, 2003, from http://www.pinchot.org/pic/Pinchot_Report_Certification_Dual_Assessment.pdf.

Mayer, D., & Hoch, D. (1993). International environmental protection and the GATT: The tuna/dolphin controversy. *American Business Law Journal, 31*(2), 187–244.

Meidinger, E. E. (2003). Forest certification as environmental law making by global civil society. In E. Meidinger, C. Elliott, & G. Oesten (Eds.), *Social and political dimensions of forest certification.* (Eifelweg, FRC: www.forstbuch.de), pp. 293–329.

Meidinger, E. E. (1999–2000). "Private" environmental regulation, human rights, and community. *Buffalo Environmental Law Journal,* 7(1). Retrieved February 6,2003, from http://www.law.buffalo.edu/homepage/eemeid/scholarship/hrec.pdf.

Meridian Institute. (2001, October). Comparative Analysis of the Forest Stewardship Council and Sustainable Forestry Initiative certification programs. Retrieved February 6,2003, from http://www2.merid.org/comparison/.

Mishra, R. (1999). *Globalization and the welfare state.* Cheltenham, UK: Edward Elgar.

Oliver, R. J. W. (1996, October 2). Progress in timber certification initiatives world wide. Forests Forever. Retrieved November 15, 2002, from www.forestsforever.org.uk/timbcert.html.

Ostrom, E. (1990). *Governing the commons—The evolution of institutions for collective action.* Cambridge, UK: Cambridge University Press.

Ruggie, J. G. (1983). Continuity and transformation in the world polity: Toward a neorealist synthesis. *World Politics, 35*(2), 261–285.

Scott, J. (1998). *Seeing like a state.* New Haven, CT: Yale University Press.

SGS Qualifor. (2004). "Forest Management Certification." Retrieved June 8, 2004, from http://www.sgsqualifor.com/fmc/htm.

Strange, S. (1996). *The retreat of the state.* Cambridge, UK: Cambridge University Press.

UN Conference on Environment and Development (UNCED). (1992). "Non-legally binding authoritative statement of principles for a global consensus on the management, conservation and sustainable development of all types of forests." A/CONF.151/26 (Vol. III), August 14. Retrieved June 8, 2004 from http://www.un.org/documents/ga/conf151/aconf15126-3annex3.htm.

United Nations Forum on Forests (UNFF). (2004). Retrieved June 8, 2004, from http://www.un.org/esa/forests/.

United Nations Forum on Forests (UNFF). (n.d. a.). IPF/IFF Process (1995–2000). Retrieved February 7, 2003, from http://www.un.org/esa/forests/ipf_iff.html.

United Nations Forum on Forests (UNFF) (n.d. b.). History and milestones of global forest policy. Retrieved February 7, 2003, from http://www.un.org/esa/forests/about-history.html.

Vogel, S. K. (1996). *Freer markets, more rules—Regulatory reform in advanced industrial countries.* Ithaca, NY: Cornell University Press.

Wiener, J. (1999). *Globalization and the harmonization of law.* London: Pinter.

Zaelke, D., Orbuch, P., & Houseman, R. F. (Eds.). (1993). *Trade and the environment—Law, economics, and policy.* Washington, DC: Island Press.

CHAPTER FOUR

The Role of ISO 14000 and the Greening of Japanese Industry

ERIC WELCH AND MIRANDA A. SCHREURS

Introduction

Is it possible to transform a national image from Japan, Inc. to Green, Inc.? At the beginning of the 1990s, Japanese corporations were widely accused for the role they played in environmental degradation. By the end of the decade, Japanese companies were leading the world in terms of the number of companies applying for international certification of their environmental management systems. This chapter addresses this puzzle. It asks why corporate Japan has been so aggressive in the past decade in pursuing environmental management certification by the International Standard Organization (ISO)? It also considers what the implications of this trend may be for corporate Japan's environmental performance.

Japanese corporate behavior in the environmental realm has been a subject of considerable interest and controversy. To date, assessments Japanese corporate environmental performance are mixed. On the one hand, there are many critics who have accused Japanese corporations of disregarding environmental protection concerns, particularly overseas. These critics have pointed to corporate Japan's heavy exploitation of natural resources in the developing world and the decision to relocate some of the most heavily polluting industries to Southeast Asia (McGill, 1992, p. 33). Japanese industries have been criticized for the role they have played in tropical deforestation (Dauvergne, 1997, 2001; Bevis, 1995) and for profiting from environmentally destructive official development aid projects in developing countries (Ensign, 1992). They also have been accused of destroying much of Japan's natural beauty and many environmentally sensitive areas in a half-century long construction boom. The construction industry has paved over Japan's coastlines, rivers, and mountainsides and pork barrel projects have resulted in many airports, leisure facilities, and golf courses being built in areas where they are hardly used (Kerr, 2001).

On the other hand, there are also many positive appraisals of corporate Japan's environmental performance. Japanese companies are among the most energy efficient in the world and are leaders in the development, manufacturing, and employment of many forms of pollution control technology (Moore and Miller, 1994). In 2001, the Organization for Economic Cooperation and Development (OECD) assessed Japan's environmental performance. It reached the following conclusion: "Overall, the mix of instruments used to implement environmental policy is highly effective. . . . Japanese industry has been proactive in establishing environmental management and reporting systems, and several branches have taken initiatives to reduce their environmental 'footprint'" (OECD, 2002, p. 22).

Given these competing visions of corporate Japan's environmental performance, this chapter considers the driving factors behind Japanese companies' rush to establish environmental management and audit systems beginning in the mid-1990s. The chapter begins with an overview of how corporate Japan has historically addressed environmental protection, focusing on the relationship between the government and industry. It then considers both the traditional importance of administrative guidance, and the use of voluntary environmental agreements in environmental implementation in Japan. The following sections explore the significance of adopting environmental management systems in Japan and compares the rates of adoption with other countries. Survey data is then used to examine why Japanese firms have turned so enthusiastically to the ISO 14001 environmental management standard. The concluding sections of the chapter explore the implications of this new trend among Japanese companies for environmental protection domestically and abroad.

Environmental policy development in Japan

There have been two periods of extensive environmental policy change in Japan that have strongly affected corporate Japan. The first was in the late 1960s and 1970s. During this period, Japanese companies were harshly criticized for their lack of attention to the environmental and human health related consequences of their activities (Huddle & Reich,

1975). Japanese companies were blamed for turning Japan into one of the most polluted countries in the world in the 1960s.

The Japanese government shared a good part of the blame for its failure to require pollution control and its compliance in hiding scientific evidence from the public that suggested a strong causal linkage between industrial pollution and human health problems. All this was done in the government's bid to help bring Japan's technological levels up to those of Western nations. The famous image of a Japan, Inc. grew out of this period of rapid economic catch-up (Johnson, 1982). The close relationship that existed among the bureaucracy, a dominant Liberal Democratic Party (LDP), and the business community made Japan's climb to the position of second richest nation in the world possible, but at tremendous cost to the natural environment and the quality of human life.

Environmental conditions became so bad that it led to a period of tremendous citizen activism. Citizens' movements protesting pollution in their communities emerged across the country. Their protests became louder and more effective over time as they gained support from lawyers, scientists, journalists, and foreign visitors. This was unusual in a society where governing had traditionally largely been left to the elite cadre of career civil servants and LDP politicians (Huddle & Reich, 1975; McKean, 1981; Broadbent, 1998). Pressure began to build on the national government to control industrial pollution. Several court decisions in favor of victims of industrial pollution placed additional pressures on the government and industry to act (Upham, 1987). As a result, in the period between 1967 and the mid-1970s, the Japanese government introduced an array of new environmental laws. These laws forced Japanese industry to install pollution control equipment and improve environmental practices. They also required the government to monitor and regulate industrial pollution. By the end of the 1970s, environmental conditions had improved noticeably.

In the 1990s, in response to growing international concern with environmental degradation of the planet, there was a second wave of extensive environmental policy activity in Japan. During this decade, the Japanese government launched many new environmental programs and instituted many new environmental laws. These included the Basic Law for Environmental Protection of 1993, the Basic Environmental Plan of 1994, the Environmental Impact Assessment Law of 1997, the Law Concerning Measures to Cope with Global Warming of 1999, and the Basic Law for Establishing a Recycling-Based Society of 2000. Japan became the world's largest provider of official development assistance in the early 1990s and began targeting more and more of this aid towards financing of environmental projects overseas. The government also began hosting international environmental conferences, setting up new environmental research think tanks, and encouraging environmental education in the schools. Japan hosted the "Third Conference of the Parties to the Framework Convention on Climate Change" where the Kyoto Protocol addressing GHE emissions was formulated (Tsuru, 1999; Schreurs, 2002).

During this period, corporate Japan suddenly began to show interest in international environmental issues as well. A large number of books began to appear discussing global environmental business strategies with titles like *Global Environmental Business*, an annual publication first produced by the Ecobusiness Network (*Ecobijinesunettowaaku*) in 1995. Firms began an array of new activities including green advertising, opening environmental offices, introducing voluntary environmental action programs, and establishing environmental management systems.

Policy instruments for environmental protection: The Japanese approach

During these two periods of extensive environmental policy activity, the Japanese government and industry developed an array of policy instruments to meet environmental protection goals. As in other industrialized countries, there was considerable use of command and control regulation in Japan. The environmental laws introduced in the early 1970s required that the most egregious polluters convert production processes, seal off effluent pipes, or shut down operations. A mix of taxes, penalties, and subsidies were used to ensure industrial compliance with environmental goals.

In addition, more flexible approaches to pollution control that made use of administrative guidance became popular. Regulations can be highly inflexible. Administrative guidance can be used as an alternative means of pursuing behavioral change. Chalmers Johnson (1982) saw administrative guidance as a critical component of what he referred to as Japan's developmental state. In applying administrative guidance, the Japanese government does not simply guide, but instead negotiates with industry regarding the establishment of rules and their enforcement (e.g., Samuels, 1987; Okimoto, 1989). Administrative guidance falls outside of the formal legal structure. This makes it attractive to government and industry alike because it allows for greater flexibility in achieving administrative goals. Yet, there are many potential problems that can emerge from heavy reliance on this kind of extralegal procedure. Administrative guidance suffers from lack of transparency and the potential for corruption. Questions also may emerge regarding the ability of government to penalize industry for noncompliance.

Nevertheless, ministries at the national level and local governments have developed a variety of incentives at their disposal to induce a firm's compliance with their administrative guidelines ranging from positive incentives like tax breaks, favorable loans, and procurement of equipment or materials, to negative incentives, such as failure to renew licenses or to permit expansion of operations (Gresser, Fujikura, & Morishima, 1981, p. 260; Imura, 2003a, 2003b). Although Japan has many laws that provide for criminal prosecution of polluters, the Japanese bureaucracy has been reluctant to turn to the courts for enforcement of regulations (OECD, 2002, p. 23). Instead, it tends to favor the use of administrative guidance to nudge and cajole industries into compliance. Industry has preferred such guidance, which is premised on long-term relationships and a high degree of mutual trust, to regulations, which if not met have the potential to result in criminal prosecution in the courts (Feinerman & Fujikura, 1998).

Voluntary pollution control and environmental agreements

One outcome of the cultural preference for administrative guidance in Japan has been the widespread establishment of voluntary pollution control and environmental agreements by industry. As early as the 1960s and 1970s, local governments began to make use of ordinances and pollution control agreements essentially requiring industries in their region to implement more stringent standards than those set nationally (Gresser, Fujikura, & Morishima, 1981, pp. 254–255). There are now tens of thousands of such voluntary agreements at the local level. Typically these voluntary agreements set emissions levels that industries must meet in a given region and mandate that industry perform at levels above and beyond those required by national legislation. Industry has some say in the shape of these agreements and there are no legal penalties for noncompliance. Yet, it is often in the industry's best interest to comply with these agreements or risk souring relations with the local government and citizens (Matsuno, 2003; Welch & Hibiki, 2002).

Voluntary agreements are also being used to address global environmental protection goals. The Japan Federation for Economic Organizations (Keidanren), a peak organization representing a broad array of industries, for example, announced a Keidanren Voluntary Environmental Action Plan in 1996. The plan urges all member industries to establish voluntary plans for GHG emissions reductions and reduction of waste generation and recycling of materials. It also urges them to introduce environmental considerations into overseas operations and to introduce environmental management and auditing systems. (Keizaidantairenjokai, 1997). The widespread adoption of environmental management systems in Japan, therefore, can be viewed as an important element of Japanese industries' voluntary approach to environmental protection.

The International Standards Organization Management Series

The ISO has established rules and procedures for obtaining management certification in a number of areas, such as its 9000 quality management series and its 14000 environmental management series. Interestingly, when the ISO first published its 9000 series of quality management systems in the early 1990s, Japanese firms were slow to react. The ISO 9000 standards require that a firm adopt policies, plans, goals, and procedures to develop a quality assurance system (Park, 1998). In the land that was known in business management schools for its quality improvement circles, this international standard did not have strong initial appeal. Perhaps this was simply because Japanese firms felt confident enough in the reputation of their products to not seek out international certification (Corbett and Russo, 2001). Alternatively, the ramifications in terms of lost contracts of not obtaining ISO 9000 certification may not have been apparent to the firms initially. Whatever the reason for the lukewarm response to ISO 9000 (Figure 4.1), in contrast, Japanese firms have been among the quickest in the world to pursue ISO 14000 certification (Figure 4.2).

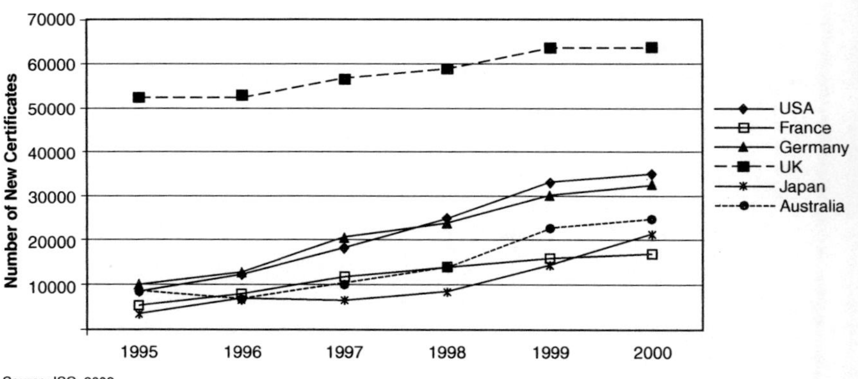

Figure 4.1. New adoptions of ISO 9000 by country, 1995–2000

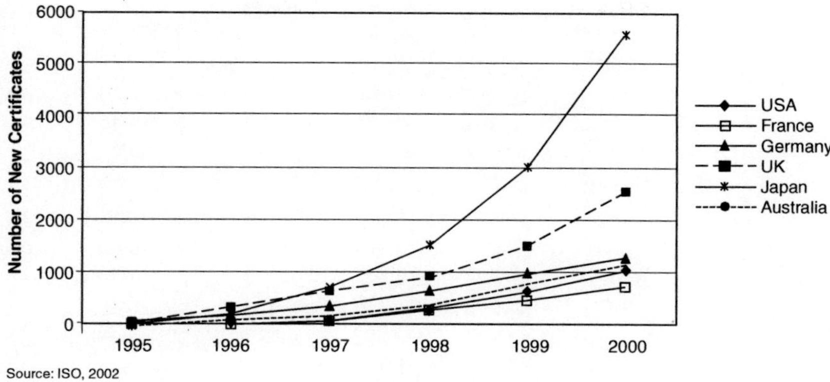

Figure 4.2. New adoptions of ISO 14001 by country, 1995–2000

ISO 14000

ISO 14000 comprises a series of voluntary environmental management standards designed for adoption by organizations of all sizes, sectors, and compositions. First published in 1996 by the ISO, the series includes the now well-known ISO 14001, a specification standard that contains five main elements designed to work together in a continuous environmental improvement cycle: (1) the environmental policy, (2) the environmental management plan, (3) plan implementation, (4) continuous monitoring, and (5) management review (Cascio, 1996; Harrington & Knight, 1999; Krut & Gleckman, 1998). The environmental policy states the commitment of the organization and forms the foundation upon which the organization establishes specific environmental targets and objectives. It requires a top management pledge to continual environmental improvement, prevention of pollution, compliance with relevant environmental regulation, and other environmental responsibilities taken on by the organization. The policy must also reasonably address the environmental impacts of organizational activities, be communicated internally, and be disseminated publicly on request.

The environmental management plan comprises three general steps: (1) identification of activities, products and services that have environmental impacts over which the organization has control, (2) determination of environmentally related legal and other requirements of the organization, and (3) establishment of environmental targets and objectives. Implementation addresses the factors of production necessary to carry out the plan. It includes evidence for the provision of necessary resources, the assignment of roles and responsibility, appropriate training, and the establishment of documentation systems.

ISO 14001 requires that organization activities having important environmental impacts be monitored on a regular basis using an environmental audit. Additionally, corrective action procedures must be established to remedy areas of noncompliance with the environmental plan. Finally, management must review the environmental management system (EMS) on a regular basis to ensure its continued relevance and effectiveness. Once an organization has satisfied these requirements, it is eligible for certification by an officially recognized ISO 14001 agency (designated at the national level). Obviously, the process of certification is potentially complex and costly, indicating that the decision to voluntarily adopt ISO 14001 is not taken lightly. Nevertheless, adoption levels of ISO 14001 continue to rise in Japan and around the world (Zharen, 1995; Lamprecht, 1997; Prakash, 1999; Mohammed, 2000).

Stepping back slightly from the descriptive detail of ISO 14001, there are five important broader elements of the standard that should be highlighted. First, the EMS is a process standard not technical standard. This means that specific targets and objectives, such as level of sulfur dioxide or carbon dioxide emissions, are set by the organization, not by the standard. Second, established regulatory standards form the environmental baseline for certification. This means that volunteering organizations can be certified if their environmental targets are set only to compliance levels, rather than the more commonly assumed beyond compliance levels. Third, there are no sanctions for failure to show improvement or, even, for failure to comply with regulations. "An organization can be registered if it is not 100 percent compliant as long as it has a system in place to identify and comply with relevant environmental regulations and it responds appropriately to incidents of noncompliance" (Harrington & Knight, 1999, p. 69). Fourth, ISO 14001 does not require an external audit to be conducted by an impartial third party organization for certification; additional audits are not required over time to ensure effectiveness of the EMS. Instead, organizations choose either to audit internally or audit by external consultant. Finally, ISO 14001 is the creation of an international industry association, which means the standard is typically governed at the national level by either an industry or trade ministry, or an industry association, but not by an environment ministry (Krut & Gleckman, 1998). In general, ISO 14001 provides organizations with maximum flexibility to both set their own technical standards and develop mechanisms to evaluate and address them, factors businesses consider important for efficient and effective environmental behavior. Despite the potential, there appears to be no guarantee that ISO 14001 certification results in environmental outcomes that are better than regulatory standards or that organizations are even complying with their own agreements.

The status of ISO 14001 in Japan, Asia, and Europe

Over the last five years, Japan has led the world in terms of the number of ISO 14001 certified organizations. As of 1999, the most recent year for which comparative data is available, Japanese organizations held more than 21 percent of the world total registrations with total numbers of registered ISO 14001 organizations outpacing the second place nation, United Kingdom, by nearly two to one (Figure 4.3). Adoption levels are much higher in Japan than in the United States, France, or Germany, although it should be noted that many European firms have instead opted for the European Union's Eco-Management and Audit Scheme (1836/93) or EMAS, which incorporates an environmental management system and places a stronger emphasis on verifiable improvement in environmental performance (Hasek, 1998).

The growth rate of ISO 14001 adoptions in Japan has been dramatic. Between December 1997 and September 2000 Japanese adoption levels jumped from 713 to 5,556, an average increase of just under 90 percent per year over three years (ISO, 2002; Japan Accreditation Board, 2003). In 2000 alone, new ISO 14001 registrants in Japan exceeded those of the United States by six to one (ISO, 2002). The absolute level of ISO 14001

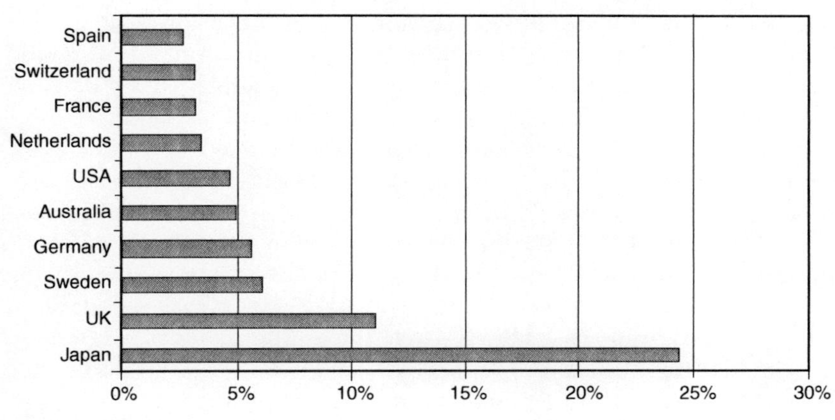

Figure 4.3. Leading national shares of global ISO 14001 certifications, 2000

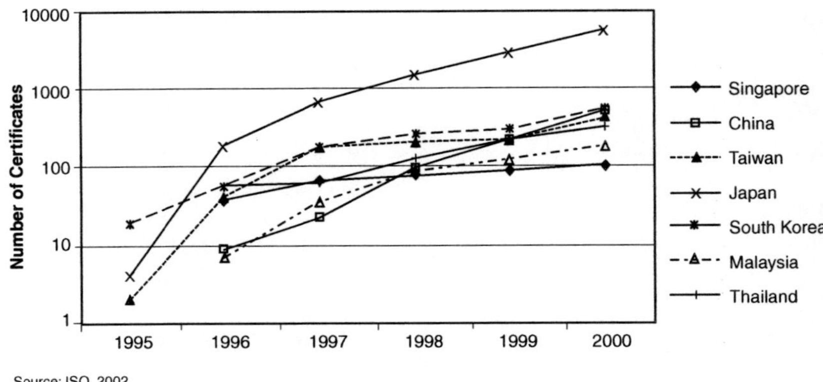

Figure 4.4. Registered ISO 14001 private sector organizations in Asian countries

registrations is higher in Japan than in any other country, and the rate of increase in registration in Japan is higher than the other top six ISO adopting nations.

A brief comparison of Japan with the rest of Asia provides an even more dramatic story. Japanese enthusiasm for ISO 14001 overwhelms adoption levels for all other Asian nations combined. Figure 4.4 shows trend lines for adoption levels in Asia over the last five years based on a logarithmic scale. While most nations hover at or below the 500 mark in 2000, Japanese adoptions represented around two thirds of all ISO 14001 volunteers in Asia. Moreover, the gap appears to be widening between Japan and South Korea, Malaysia, Singapore, and Thailand, while it has narrowed only slightly between Japan and Taiwan and China. Further, except for Taiwan and South Korea between 1999 and 2000, adoption rates have generally declined or held steady over the past four years in Asia as a whole. This finding is somewhat contrary to expectations. For example, in 1995, the Chinese government declared its official support for ISO 14001, citing future competitive advantage as an important justification for adoption (Hasek, 1998). However, adoption levels in China remain low (510 in 2000). Additionally, it is often expected that high levels of adoption of one country would help entice business partners in other nations to become certified. For example, Japanese companies might demand that their business partners in other Asian nations become certified. Nevertheless, the enthusiastic adoption levels seen in Japan appear not to have influenced strongly firms in other Asian countries. This could be the result of any number of factors, including limited national infrastructure for and communication about certification programs, limited capacity of companies to pursue the rigorous application procedure, or poor financial situations of firms in the region as they recover from the financial crisis. Firm size may also matter since the application process may be too expensive for the many small and medium-sized firms found in developing Asian states (Quazi, 1999; Hamner & del Rosario, 1997). Overall, evidence clearly indicates that Japanese adoption rates of ISO 14001 are above world and regional norms, making questions about motivation and effectiveness all the more valid.

Voluntary self-regulation: Motivators and effectiveness

It is generally recognized that voluntarism incurs costs and provides benefits to the regulated organization and to the regulator. For the organization, adoption of and compliance with voluntary standards requires an investment of time and resources. In return, the organization may benefit from any number of factors that either directly or indirectly affect their bottom line. Benefits to ISO 14001 specifically identified in the literature include more confident regulatory compliance by business, greater market access, regulatory incentives, insurance benefits, improved production efficiency, improved environmental performance, and better customer, supplier, employee, and external stakeholder relations (Harrington & Knight, 1999; ISO, 1998). From the regulator's perspective, voluntary programs are thought to provide environmental benefits (beyond what would exist without the program) without incurring accompanying high administrative costs. These benefit and cost assumptions have generally driven the popularity of voluntary programs throughout the 1990s (see chapter 5 in this volume).

Theoretical explanations of voluntarism essentially mirror the conventional wisdom on drivers of adoption and can be separated into two broad groups: market response theory and regulatory influence theory. In addition, the institutional context (laws, norms, expectations, etc.) can provide incentives, opportunities, and constraints that drive organizational behavior toward certain trajectories (Kollman & Prakash, 2002). Certain national contexts that favor a specific firm level behavioral response or policy direction may result in patterns of company behavior not evident elsewhere. For example, normative goals supporting global environmental leadership or cooperative business-government relations may be important institutional factors that facilitate voluntarism in Japan.

Regulatory influence theory

Based heavily on political economic theory of interest group pressure (Peltzman, 1976; Becker, 1983; Stigler, 1971), regulatory influence theory asserts that voluntarism indicates both an announced intention to invest in pollution control and an attempt to influence or manipulate the regulatory system. The behavioral mechanism is relatively simple. By volunteering, firms appear to be greener—more willing to over comply with environmental regulations—giving government and other stakeholders the impression that the volunteering firm poses a relatively low environmental threat. As a result, the firm is able to limit or weaken the lobbying effectiveness of citizens and environmental organizations seeking stronger regulations. According to this theory, volunteering firms may be seeking to preempt or weaken future regulations (Lyon & Maxwell, 1999; Maxwell & Decker, 1998; Lutz, Lyon, & Maxwell, 1998), or lighten existing regulatory pressure felt by the firm (Decker, 1998; Maxwell & Decker, 1998).

A version of this theory shows that the terms of the agreement and the level of compliance are determined by the allocation of bargaining power between regulator and the firm and the level of background threat of sanctions (Segersen & Miceli, 1998). Greater regulator bargaining power is expected to lead to more stringent agreements, while compliance will be greater when the regulated firm believes that nonfulfillment will trigger regulatory action. Recent research supports the expectation that compliance can be low when the background threat is also low (Welch, Mazur, & Bretschneider, 2000; King & Lenox, 2000). Regulatory influence theory has two specific implications for ISO 14001 adoption in Japan. First, it predicts that firms adopt ISO 14001 to influence regulatory processes in their favor. Further, it implies that more heavily regulated firms would be more likely to volunteer as they would have the most to gain from a cleaner, more cooperative image. Second, regulatory influence theory predicts that the low background threat and regulator bargaining power would reduce the effectiveness of ISO 14001 implementation.

Despite this clear line of logic, regulatory influence theory assumes the institutional context is adversarial and legalistic. In such an environment cooperative interaction, requiring extensive exchange of information and trust, is suspect either because of fears of regulatory capture (Kollman & Prakash, 2002) or because information shared can be turned against firms at some later point (Lamphrect, 1997 as cited in Quazi, 1999). By contrast, the Japanese institutional context is seldom described as adversarial. Instead, a cooperative context exists in which government and business share authority in the regulatory and market domains (Samuels, 1987; Okimoto, 1989; Welch & Hibiki, 2002). The cooperative context produces policy processes such as market-oriented administrative guidance and self-regulatory voluntary environmental agreements. In a cooperative regulatory context, legal concerns about information sharing may be lower, expectations of cooperative behavior higher, and efforts to influence the regulatory context more subtle than in adversarial regimes. Moreover, violation of agreements or understanding by either side carries with it broader legal consequences; violation implies willful breach of a social contract. Therefore, evidence indicating the government explicitly favors or promotes ISO 14001, or that the private sector downplays regulatory advantages but emphasizes broader social and environmental objectives, may help explain high adoption rates in Japan.

Market response theory

A second theoretical trajectory that can be called market response strategy assumes firms invest in environmental activities due to demand from green consumers or investors, to take advantage of potential cost savings, or to ensure access to supplier networks. Private sector organizations volunteer to satisfy the environmental requirements of consumers who either demand environmentally benign products or demand greater environmental action on the part of producers (Arora & Cason, 1996; Williams, Medhurst, & Drew, 1993). Firms may also volunteer as a means of attracting investors who may be seeking greener companies for any number of reasons including ethical considerations (Baron, 1996), the promise of future market opportunities (Hamilton, 1995; Khanna, Quimio, & Bojilova, 1998), or because strong environmental practices may indicate a lower investment risk (Khanna & Damon, 1999). Market response theory predicts highly visible consumer product and service firms (such as those in the consumer electronics industry), in regions populated by environmentally aware consumers, would be more likely to adopt ISO 14001 than less visible intermediate supplier firms. It also predicts consumer and capital market benefits would help drive voluntary adoption of ISO 14001.

Firms may also voluntarily adopt an environmental management system based on the potential for identifying cost saving strategies (Lyon & Maxwell, 1999; Buchholz, 1993; Groenewegen, Fischer, Jenkins, & Schot, 1996; Smart, 1992). For example, environmental management systems requiring environmental planning, record keeping and auditing could help identify immediate opportunities for resource or energy savings, and such savings may result in competitive advantage gains by adopters (Starik & Marcus, 2000; Hart, 1995). This low hanging fruit argument represents a short-term immediate benefit rather than a long-term continuous benefit.

Finally, as nodes within a supplier network, firms strive to comply with the norms and demands of upstream purchasers. Firms conform as a means of survival; they realize greater access to resources and stability as they adapt to industry's structures and demands (DiMaggio, 1988; Pfeffer & Salancik, 1978). Therefore, diffusion of ISO 14001 should be affected by the extent to which manufacturers expect upstream companies to prefer suppliers who are ISO 14001 certified. This effect may be especially strong in the comparatively tightly linked and highly competitive network of firms that exists in Japan.

The National Institute for Environmental Studies' Survey of Firms and ISO 14001

This chapter makes use of survey data to help determine the drivers behind high ISO 14001 adoption rates in Japan. The same data also provides some indication of the effect that ISO 14001 has on environmental behavior of firms. The National Institute for Environmental Studies (NIES)—the research arm of Japan's Environment Agency—has conducted surveys on private sector adoption of ISO 14001 in Japan. In this analysis, we use survey data gathered in March 1999 by the NIES. Questionnaires were sent to 2,918 Japanese facilities in four industries: chemical manufacturing, electronics, electrical machinery, and electric power production. These industries were chosen because each represents a significantly different market and produces different types of outputs, and because they are the leading adopters of ISO 14001 in Japan. Because ISO 14001 is a site-based program in which establishments (not companies) are certified, questionnaires were sent to facility managers.

The sample, taken from the 1996 Census of Manufacturers, was stratified into ISO 14001 adopters and nonadopters. Questionnaires were sent to all ISO 14001 certified companies in the four industries. A larger random sample of facilities was selected from the list of nonadopter facilities. Of the 718 certified facilities surveyed, 364 responded with useable data (50.7 percent), while only 445 of the 2,200 non-ISO facilities surveyed provided complete responses (20.2 percent). Although a low response rate from nonadopters was anticipated, a figure closer to 30 percent was expected. Responses were entered independently into two separate databases and subsequently cross-checked for errors. The low non-ISO response rate may result in response bias as larger firms were more likely to respond to the survey. However, the size distribution of adopters and nonadopters was comparable, providing a potentially more meaningful and robust analysis of large-sized ISO and non-ISO facilities.

What explains Japan's high adoption levels?

Table 4.1 displays results from the survey question asked of ISO 14001 certified facilities: What are the advantages of acquiring certification for your establishment? Possible answers include improved corporate image, better facility managerial systems, maintenance and expansion of overseas customers, maintenance and expansion of domestic customers, reduction of environmental load, improvement of products and services, avoidance of accidents, disasters, and other risks, improvement of employee morale, cost reduction, improvement of financing and fund raising, future preferential treatment by administrative authorities, and maintenance of favorable relationship with citizens and society. The table reports proportions of facilities in three industries that responded affirmatively to the question. Because facilities in the electric power

Table 4.1. Proportion of facilities indicating advantages of ISO 14001 certification

	Electric Machinery	Chemical Manufacturing	Electronics
Improved image	0.62	0.33	0.45
Improved general management	0.50	0.27	0.32
Better overseas competitiveness	0.37	0.14	0.19
Better domestic competitiveness	0.40	0.14	0.19
Reduce environmental load	0.58	0.31	0.45
Product improvement	0.33	0.11	0.19
Risk avoidance	0.48	0.22	0.29
Employee morale	0.53	0.28	0.41
Cost reduction	0.33	0.20	0.19
Financing advantages	0.04	0.01	0.01
Preferential government treatment	0.05	0.03	0.01
Improved consumer relations	0.46	0.19	0.34

industry often requested the parent company to respond to the questionnaire, electric power responses were dropped from the sample.)

In general, a high proportion of facilities indicated that image, management, environmental impact, risk avoidance, and employee morale were the most important considerations for ISO 14001 adoptions. Cost reduction, product improvement, customer relations, and competitiveness factors represented a middle level of considerations across the industries, while financing and government elements received the lowest proportion of affirmative responses. These findings show image plays a central role in adoption of ISO 14001, and indicates Japanese firms are seeking both to present the international community and domestic consumers with a greener image and to improve their environmental performance. Moreover, environmental, image, and managerial factors are more important than perceived economic advantages, although as a group, these competitiveness factors are clearly important. The fact that few facilities believe ISO 14001 will help relationships with government—a finding that is supported in other recent research (Mohammed, 2000; Welch, Rana, & Mori, forthcoming)—may be attributable to the cooperative regulatory context within which facilities operate. Adoption is not undertaken to improve regulatory relationships, but rather as fulfillment of an expectation or as the result of administrative guidance.

When facilities were asked about the extent to which they value the reactions from stakeholders concerning their environmental efforts and actions, ISO adopters responded with much higher assessments of the value of all groups, compared to nonadopters. These differences were statistically significant at the 0.05 level in all cases. Although the rankings of the different stakeholders for the two subsamples are slightly different, the top four stakeholders for both adopters and nonadopters are local government, regional society, and domestic consumers. Central government is also fourth in the ISO 14001 subsample, perhaps indicating the importance of central government guidance with respect to ISO 14001. Overall, the findings in Table 4.2 fill in some detail missing from Table 4.1: probable targets of image improvement provided by ISO 14001 includes, within the top four, local government, civil society, and business. It is interesting to note Central Government is fourth on the list while Local Government is first. This may indicate ISO 14001 diffusion has benefited from the long-term institutionalization and trust of local voluntary environmental agreements as a means of effecting cooperative environmental policy in Japan.

Beyond this data, environmental laws and elements of Japan's public sector have provided support and encouragement for firms to adopt ISO 14001. The Basic Environmental Law sets the stage for

Table 4.2. Value of stakeholder reactions for environmental activity

	ISO 14001 Nonadopters	ISO 14001 Adopters
Local government	3.56 (1.10)	4.28 (0.82)
Regional citizens and organizations	3.61 (1.05)	4.13 (0.85)
Domestic consumers	3.50 (1.07)	4.08 (0.82)
Central government	3.41 (1.09)	3.99 (0.93)
International consumers	3.00 (1.21)	3.96 (0.90)
Employees	3.44 (0.98)	3.92 (0.82)
Stockholders	3.01 (1.67)	3.88 (0.92)
Media	3.17 (1.10)	3.82 (0.84)
Experts	3.07 (1.05)	3.58 (0.89)

specific language in the Basic Environmental Plan (1994) that supports voluntary adoption of environmental management systems including ISO 14001:

> [This plan seeks] to promote environmental management, which consists of establishment of corporate policies on environmental conservation, target setting, planning, organizational arrangements such as the appointment of managers, and audit of the system. In doing this, discussion at the International Standards Organization should be taken into account. (Basic Environmental Plan, chap. 3, sec. 1.3, 1994)

To facilitate and streamline ISO 14001 adoption, the Ministry of International Trade and Industry (MITI), which was renamed the Ministry of Economy, Trade, and Industry (METI) in 2001, has designated the Japan Accreditation Board (JAB) as the caretaker of ISO 14001. JAB established conformity between ISO 14001 standards and Japan Industry Standards (JIS) such that all ISO 14001 standards have complementary JIS. Since 1996, JAB has also overseen a centralized, reliable organization structure that conducts accreditation and registration of all ISO registrars, certifiers, and auditors in Japan. At the time of this writing JAB had accredited twenty-seven registration bodies, sixteen auditor training bodies and one auditor certification body (JAB, 2003). These ISO bodies form the primary source of information for adopter firms (Mohammed, 2000). The clear and streamlined structure for ISO 14001 implementation may contribute to relatively low cost channels through which organizations can gain information, assistance, and accreditation.

While no sanctions are associated with nonadoption and no direct benefits or pressure for adoption are provided by METI or JAB, certain other elements of the institutional structure in Japan facilitate ISO adoption. For example, some local governments provide financial assistance to organizations seeking to register for ISO: the Tokyo Metropolitan government covers half of the registration costs up to a maximum of 1.3 million yen (about $10,000) (Standards Council of Canada [SCC], 1999). In addition, many of the largest companies in Japan have registered most or all of their facilities worldwide for ISO 14001, and some either give priority to suppliers who are ISO 14001 certified or require registration outright before contracts can be renewed (SCC, 1999; Hasek, 1998).

What are the effects of ISO 14001 adoption?

Using ISO 14001 survey data, we further address two questions concerning the effects of ISO 14001 adoption: (1) How are internal management processes and structures of firms affected by ISO 14001 adoption? and (2) How does ISO 14001 affect environmental outcomes? This latter question responds to the most frequent criticism of ISO 14001, which is that the standard may not actually result in improved environmental quality. These criticisms arise partly from the fact that ISO 14001 does not set specific performance criteria and partly because analysts have shown improvement in environmental performance often results from employee involvement, rather than from top-down implementation of an environmental management system (Boiral & Sala, 1998, pp. 61–62).

Figure 4.5 displays results from a survey question on the type and level of environmental management change undertaken by the organization as part of the ISO 14001 certification process. In this figure, solid black portions of the graph represent companies able to use existing environmental management structures and processes—no changes were needed for ISO certification. Shaded portions indicate existing structures and processes that were modified for ISO 14001, and solid white portions represent entirely new structures and processes developed by the organization for certification. Each column represents a specific element of the ISO 14001 EMS: environmental policy (Policy), identi-

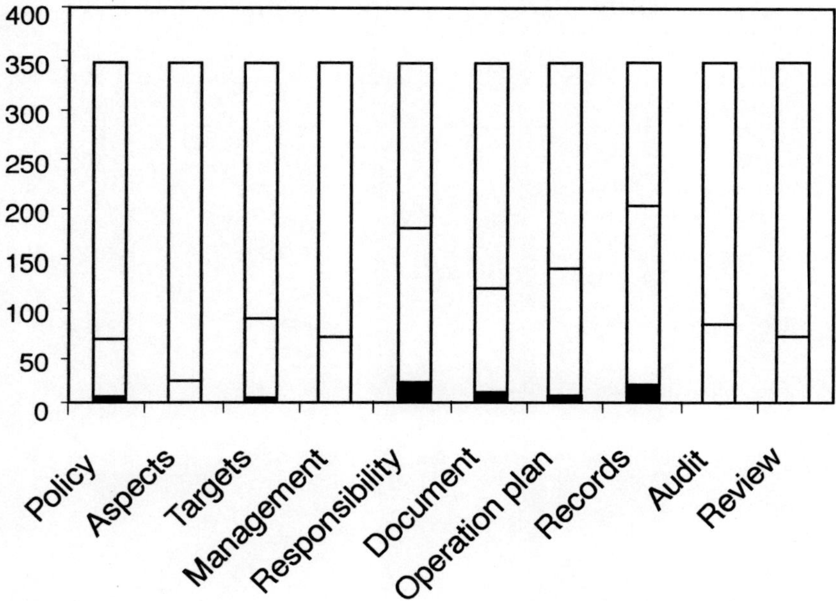

Figure 4.5. Management changes undertaken by ISO 14001 adopters in Japan

fication of environmental aspects and their influences (Aspects), objectives and targets (Targets), environmental management programs (Management), structure and responsibility (Responsibility), control system documentation (Document), operational control plan (Operation Plan), environmental records maintenance (Records), environmental audit (Audit) and management review process (Review).

For most columns, the majority of organizations developed new management structures and processes to comply with ISO 14001 requirements. Two noted exceptions are Responsibility and Document wherein a slim majority has revised existing plans rather than developed new ones. Nevertheless, overall it appears ISO 14001 certification results in some internal environmental management changes in adopter organizations.

When facilities were asked about a range of environmental activity, ISO firms indicated more active environmental performance than

Table 4.3. ISO and non-ISO environmental activities (N, mean, standard deviation)

Environmental activity	ISO	Non-ISO
Restructure business relationship	318* 1.21 (4.14)	370 0.49 (2.21)
Renegotiate business contract	339 0.24 (1.71)	386 0.18 (1.13)
Changed supplier partners	342* 0.11 (0.71)	121 0.01 (0.14)
Frequency internal environmental reports	346* 3.99 (1.25)	384 2.69 (1.62)
Extent environment in business decisions	345* 3.64 (0.68)	381 2.60 (1.20)

*Statistically different means at $p < 0.05$

non-ISO firms. Table 4.3 shows results for five different questions. The first and second questions report the number of times the facility has restructured a business relationship and renegotiated a business contract with an existing business partner to improve environmental performance of products and processes during the past five years. The third question asked about the number of times during the past five years that the facility had changed a business partner for environmental reasons. While overall numbers are rather low, there are significant differences between ISO and non-ISO firms for two of the three activities. This clearly indicates that ISO firms are fundamentally more environmentally active. However, it does not indicate whether or not ISO adoption has had any effect; that is, causes firms to become more environmental. It is entirely possible, for example, that ISO adopters represent that portion of the population that is more environmentally active in the first place.

The last two rows of Table 4.4 report results from two questions on the frequency at which the facility creates internal environmental reports for review of progress toward environmental targets and the extent to which environmental factors are considered when making business decisions (five point scale, low to high). Findings continue to support the contention that ISO firms are more environmental than non-ISO firms, but provide little causal evidence of the effect of ISO on behavior.

Although the inability to address questions of effectiveness is a particular problem of survey questionnaires, it is possible to separate

Table 4.4. Evaluation and monitoring behavior by year of ISO certification

Certification year	Frequency of evaluation of environmental target progress	Frequency of monitoring environmental targets
1999	3.74 (1.38)	2.21 (0.90)
1998	4.31 (1.17)	2.59 (0.92)
1997	4.26 (1.20)	2.60 (0.94)
1996	4.14 (1.37)	2.65 (0.89)
1995	4.14 (1.21)	2.71 (0.95)

out levels of environmental activity according to the year in which a facility attained certification. If, as expected by many, ISO 14001 drives continuous environmental change in organizations, it may be possible to see that earlier adopters are more active than recent volunteers. However, the data in Table 4.4 are generally not supportive of ISO as a change agent. Responses to two questions about the frequency of evaluation of progress toward targets and the frequency of monitoring of targets, both important elements of ISO, do not appear to change dramatically over time. Monitoring is measured on a four-point scale (continuous, weekly, monthly, and annually), while evaluation is measured on a five-point scale (greater than 6, 5 to 6, 3 to 4, 2, and 1 time(s) per year), where a lower number represents a higher frequency. While the most recent adopters (1999) report a higher frequency of evaluation and monitoring, this difference quickly plateaus at a lower frequency level for earlier adopters. While adopter frequencies are higher than nonadopter frequencies (not reported here), the important point is that little change occurs over time. Research elsewhere has found those firms undergoing certification are significantly more environmental than nonadopters but significantly less environmental than certified firms. Combined with the evidence here, it may be that ISO 14001 compels facilities to conform to a certain environmental standard, however the level of ongoing environmental performance improvement may be limited (Welch, Rana, & Mori, forthcoming).

Institutional advantage, environmental image, and ISO 14001 certification

There appears to be two factors working together to produce the significant levels and rates of adoption of ISO 14001 in Japan. First, the Japanese institutional setting facilitates information dissemination and provides assistance to private sector ISO 14001 adopters. The context within which Japanese enterprises operate provides incentives for adopters and enhances the ability of firms to adopt ISO 14001. Additionally, there seems to be an increasing impression that greenness, or the image of greenness, pays.

For example, Fujitsu Ltd., a major electronics company, released Japan's first official environmental auditing report claiming that the firm had reaped a net gain of 4 billion yen ($33 million) in 1998 as a result of its environmentally friendly expenditures. This figure was derived using environmental accounting guidelines produced by Japan's Environment Agency. For FY 99, the firm estimated economic benefits from environmental spending would bring the firm 11 billion yen ($90 million). Sony Corporation was also expected to release a similar report that would divide the firm's environmental expenditures into several categories, including investment in pollution control, expenditure to cut energy consumption, and the expense of ISO 14001 certification (Kakuchi, 1999).

The resulting picture is (at least) a partially familiar one: government and the corporate sector cooperatively promoting a flexible policy solution from which each benefits. Domestic consumers show their preferences through the marketplace and trading partners take notice as Japan seeks to establish environmental visibility. Whether or not environmental benefits actually result from the implementation of an environmental management system, there is clearly a positive image associated with adoption of ISO 14001 in Japan. Moreover, the effects of Japan's ISO 14001 promotion and adoption efforts are beginning to be felt elsewhere.

It is becoming more common for Japanese multinationals to have their foreign affiliates seek ISO 14001 certification. This trend is particularly visible among the auto and electronics industries. In many cases, foreign affiliates are given several years of lag time to follow the lead in obtaining certification set by their corporate parent in Japan. For example, Toyota Motor Corporation, which buys about 80 percent of its parts from outside companies, urged all its suppliers to obtain ISO 14001 certification by 2003 (Asia Pulse, March 24, 1999). Matsushita Electric, manufacturer of the National and Panasonic brands, set the end of fiscal 1999 as a target date for all of its manufacturers (138 manufacturing locations, 80 domestic and 58 overseas) to obtain ISO 14001 certification (Matsushita Electric Works Ltd., 1998).

In the long-term, the hope is this trend will indeed have a positive effect on the environmental performance of Japanese firms and their suppliers. Voluntary compliance clearly has its limitations, but it is clear this has become an important component of how Japanese society goes about environmental protection. To the extent voluntary approaches to environmental protection are relied upon in Japan and overseas, it is important to establish a means of assuring compliance with voluntary goals. One possible mechanism for compliance may be found in the informal institutions that have emerged from Japan's experience with local voluntary pollution control agreements.

References

Arora, S., & Cason, T. (1996). Why do firms volunteer to exceed environmental regulations? Understanding participation in the EPA's 33/50 Program. *Land Economics, 72,* 413–432.

Baron, D. (1996). *Business and its environment.* Upper Saddle River, NJ: Prentice Hall.

Basic Environmental Plan (Kankyo Kihon Keikaku). (1994). Tokyo: Government Printing Office, Ministry of Finance.

Becker, G. (1983). A theory of competition among pressure groups for political influence. *Quarterly Journal of Economics, 98,* 371–400.

Bevis, W. (1995). *Borneo log: The struggle for Sarawak's forests.* Seattle: University of Washington Press.

Boiral, O., & Sala, J. (1998). Environmental management: Should industry adopt ISO 14001? *Business Horizons, 41*(1), 57–65.

Broadbent, J. (1998). *Environmental politics in Japan: Networks of power and protest.* Cambridge, UK: Cambridge University Press.

Buchholz, R. (1993). *Principles of environmental management: The greening of business.* Englewood Cliffs, NJ: Prentice Hall.

Cascio, J. (Ed.). (1996). *The ISO 14000 Handbook.* Milwaukee, WI: ASQ Quality Press.

Corbett, C. J., & Russo, M. V. (2001, December). ISO 14001: Irrelevant or invaluable?. *ISO Management Systems,* 23–29.

Dauvergne, P. (1997). *Shadows in the forest: Japan and the politics of timber in southeast Asia.* Cambridge, MA: MIT Press.

Dauvergne, P. (2001). *Loggers and degradation in the Asia Pacific.* Cambridge, MA: Cambridge University Press.

Decker, C. S. (1998). Implications of regulatory responsiveness to corporate environmental compliance strategies. Working Paper, Indiana University, Department of Business Economics and Public Policy, Kelley School of Business.

DiMaggio, P. (1988). Interest and agency in institutional theory. In L. G. Zucker. (Ed.), *Institutional patterns and organizations: Culture and environment* (pp. 3–21). Cambridge, MA: Ballinger.

Ecobijinesunettowaaku, ed. (1997). Chikyû kankyô bijinesu (global environmental business). Tokyo: Office May Corporation.

Ensign, M. (1992). *Doing good or doing well: Japan's foreign aid program.* New York: Columbia University Press.

Feinerman, J. V., & Fujikura, K. (1998). Japan's consensus-based compliance. In E. Brown Weiss & H. K. Jacobson (Eds.), *Engaging countries: Strengthening compliance with international environmental accords* (pp. 253–290). Cambridge, MA: MIT Press.

Gresser, J., Fujikura, K., & Morishima, A. (1981). *Environmental law in Japan.* Cambridge, MA: MIT Press.

Groenewegen, P., Fischer, K., Jenkins, E. G., & Schot, J. (1996). *The greening of industry resource guide and bibliography.* Washington, DC: Island Press.

Hamilton, J. T. (1995). Pollution as news: Media and stock market reactions to the toxics release inventory data. *Journal of Environmental Economics and Management, 28,* 98–113.

Hamner, B., & del Rosario, T. (1997). Green purchasing: A global channel for improving the environmental performance of small enterprises. Background Paper No. 2 for the Workshop on Environmental Policy: Globalization and the Environment: New Challenges for the Public and Private Sectors, November 13 and 14, SG/DNME/EPOC(97)5. Paris: OECD.

Harrington, H. J., & Knight, A. (1999). *ISO 14000 implementation: Upgrading your EMS effectively.* New York: McGraw-Hill.

Hart, S. L. (1995). A natural resource based view of the firm. *Academy of Management Review, 20*(4), 986–1014.

Hasek, G. (1998). ISO's green standard takes root. *Industry Week, 247*(4), 39–42.

Huddle, N., & Reich, M., with Stiskin, N. (1975). *Island of dreams: Environmental crisis in Japan.* New York: Autumn Press.

Imura, H. (2003a). Japanese environmental policy: Financial mechanisms. World Bank Paper.

Imura, H. (2003b). Japan's environmental policy: Policy instruments. World Bank Paper.

International Standards Organization. (1998). ISO 14000—Meet the Whole Family! Geneva, Switzerland: Author.

International Standards Organization. (2002). *The ISO Survey of ISO 9000 and ISO 14000 Certificates—Tenth Cycle.* Geneva, Switzerland: Author.

Japan Accreditation Board. (2003). The Japan Accreditation Board for Conformity Assessment. Retrieved from http://www.jab.or.jp.

Johnson, C. (1982). *MITI and the Japanese miracle.* Stanford, CA: Stanford University Press.

Kakuchi, S. (1999, June 4). Environment-Japan: Going green is good for the bottom line. *Interpress Service.*

Keizaidantairengokai. (1997). Keidanren Kankyô Jishu Kôdô Keikaku. Tokyo: Keidanren Sangyô Honbu Chikyû Kankyô Guruupu.

Kerr, A. (2001). *Dogs and demons: Tales from the dark side of Japan.* New York: Hill and Wang.

Khanna, M., & Damon, L. A. (1999). EPA's Voluntary 33/50 Program: Impact on toxic releases and economic performance of firms. *Journal of Environmental Economics and Management, 37,* 1–25.

Khanna, M., Quimio, W., & Bojilova, D. (1998). Toxics release information: A policy tool for environmental protection. *Journal of Environmental Economics and Management, 36,* 243–266.

King, A., & Lenox, M. J. (2000). Industry self-regulation without sanction: The chemical industry's responsible care program. *The Academy of Management Journal, 43*(4), 698–716.

Kollman, K., & Prakash, A. (2002). EMS-based environmental regimes as club goods: Examining variations in firm-level adoption of ISO 14001 and EMAS in the U.K., U.S. and Germany. *Policy Sciences, 35*, 43–67.

Krut, R., & Gleckman, H. (1998). *ISO 14001: A missed opportunity for sustainable global industrial development.* London: Earthscan.

Lamprecht, J. (1997). *ISO 14000 Issues and implementation guidelines for responsible environemtnal management* (pp. 43–88). New York: Amcom.

Lutz, S., Lyon T., & Maxwell, J. (1998). Strategic quality choice with minimum quality standards. Discussion Paper #1793. London: Center for Economic Policy Research.

Lyon, T., & Maxwell, J. (1999). Voluntary approaches to environmental regulation: A survey. Working Paper, Indiana University, Department of Business, Economics, and Public Policy, Kelley School of Business.

Matsuno Yu (2003). Local Government and Industries: Pollution Control Agreements. World Bank Paper.

Matsushita Electric Works, Ltd. (1998). Annual Report. http://www.mew.co.jp/corp/ir/financial/annual/ar981.

Maxwell, J., & Decker, C. (1998). Voluntary environmental investment and regulatory flexibility. Working Paper, Indiana University, Department of Business Economics and Public Policy, Kelley School of Business.

Maxwell, J., Lyon, T., & Hackett, S. (1998). Self-regulation and social welfare: The political economy of corporate environmentalism. (Nota di Lavoro 55.98). Milan, Italy: Fondazione Eni Enrico Mattei.

McGill, D. (1992, October 4). Scour technology's stain with technology. *The New York Times Magazine* 6, pp. 32–37.

McKean, M. (1981). *Environmental protest and citizen politics in Japan.* Berkeley: University of California Press.

Mohammed, M. (2000). The ISO 14001 EMS implementation process and its implications: A case study of central Japan. *Environmental Management, 25*(2), 177–188.

Moore, C., & Miller, A. (1994). *Green gold: Japan, Germany, the United States and the race for environmental technology.* Boston: Beacon Press.

Okimoto, D. (1989). *Between MITI and the market: Japanese industrial policy for high technology.* Stanford, CA: Stanford University Press.

Organization for Economic Cooperation and Development. (2002). *Environmental performance reviews: Japan.* Paris: Author.

Park, J. (1998). Ecological stewardship in Japanese firms. *Corporate Environmental Strategy, 5*(3), 33–39.

Peltzman, S. (1976). Toward a more general theory of regulation. *Journal of Law and Economics, 19*, 211–248.

Pfeffer, J., & Salancik, G. R. (1978). *The external control of organizations: A resource dependence perspective.* New York: Harper & Row.

Prakash, A. (1999). A new-institutionalist perspective on ISO 14001 and responsible care. *Business Strategy and the Environment 8*, 322–335.

Quazi, H. A. (1999). Implementation of an environmental management system: The experience of companies operating in Singapore. *Industrial Management and Data Systems 7,* 302–311.

Samuels, R. J. (1987). *The business of the Japanese State: Energy markets in comparative and historical perspective.* Ithaca, NY: Cornell University Press.

Schreurs, M. A. (2002). *Environmental politics in Japan, Germany, and the United States.* Cambridge, UK: Cambridge University Press.

Segersen, K., & Miceli, T. (1998). Voluntary approaches to environmental protection: The role of legislative threats. *Journal of Environmental Economics and Management, 36,* 109–130.

Smart, B. (1992). *Beyond compliance: A new industry view of the environment.* Washington, DC: World Resources Institute.

Standards Council of Canada. (1999). ISO 14000 in Japan. *Consensus Magazine, 26*(5), 14–15.

Starik, M., & Marcus, A. A. (2000). Special research forum on the management of organizations in the natural environment: A field emerging from multiple paths, with many challenges ahead. *The Academy of Management Journal, 43*(4), 539–547.

Stigler, G. (1971). The theory of economic regulation. *Bell Journal of Economics and Management Science, 2,* 3–21.

"Toyota Draws Up Green Guidelines for Suppliers," Asia Pulse, March 24, 1999.

Tsuru, S. (1999). *The political economy of the environment: The case of Japan.* London: Athlone Press.

Upham, F. (1987). *Law and social change in postwar Japan.* Cambridge, MA: Harvard University Press.

Welch, E. W., Mazur, A., & Bretschneider, S. (2000). Voluntary behavior of electric utilities: Contribution of the climate challenge program to the reduction of CO_2. *Journal of Public Policy and Management, 19*(3), 407–425.

Welch, E. W., & Hibiki, A. (2002). Japanese voluntary environmental agreements: Bargaining power and reciprocity as contributors to effectiveness. *Policy Sciences, 35*(4), 401–424.

Welch, E. W., Rana, A., & Mori, Y. (in press). Promises and pitfalls of ISO 14001 for competitiveness and sustainability: A comparison of Japan and the United States. *Greener Management International: The Journal of Corporate Environmental Strategy and Practice,* 44, forthcoming 2004.

Williams, H., Medhurst, J., & Drew, K. (1993). Corporate strategies for a sustainable future. In K. Fischer & J. Schot (Eds.), *Environmental Strategies for Industry: International Perspectives on Research Needs and Policy Implications* (pp. 117–146). Washington, DC: Island Press.

Zharen, W. M. (1995). *ISO 14000, Understanding the environmental standards.* (pp. 39–49). Rockville, MD: Government Institutes Inc.

CHAPTER FIVE

Voluntary Agreements

Cornerstone or Fig Leaf in German Climate Change Policy?

MICHAEL T. HATCH

In the Climate Change Convention finalized at the 1992 Earth Summit in Rio, the advanced industrial nations of the world agreed to pursue the goal of stabilizing greenhouse gas emissions by the year 2000. The German government, for its part, had committed to reducing CO_2 emissions 25–30 percent by the year 2005. Efforts to put in place the set of measures necessary to achieve such an ambitious target, however, encountered several hurdles, not the least of which was German industry's resistance to regulatory measures viewed as adversely impacting its competitiveness.

In an attempt to overcome this political impasse, a voluntary agreement between the federal government and industry was announced in March 1995, just prior to the first meeting of signatories to the convention in Berlin. In this agreement, fifteen industry associations (later expanded to nineteen)—representing 80 percent of industrial energy consumption—committed to specific reductions in CO_2 emissions or specific energy consumption of up to 20 percent by the year 2005. For its part, the German government agreed to hold in abeyance additional regulatory measures designed to reduce CO_2 emissions.

This series of events reflects a more general phenomenon common to most industrial democracies in recent years: while environmental concerns remain politically salient domestically, efforts to address them have often stalled in the face of competing concerns about economic growth and international competitiveness. An added dimension is the increasing globalization of economic activity, making more explicit the link between continued economic prosperity and the maintenance of international competitiveness. These developments have heightened concerns about environmental regulation and its impact on competitiveness. As a consequence, attempts to apply the more traditional forms of environmental regulation have met ever greater opposition from the business community. Increasingly, governments and industry have entered into a dialogue resulting in voluntary agreements as an alternative to so-called command and control regulatory measures.

Several reviews of the literature on voluntary agreements reveal how little is actually known about the effectiveness of this policy instrument (EEA, 1997; Harrison, 1999). By analyzing the voluntary agreement concluded between the German government and the electrical power generation industry, this chapter provides a case study that seeks to help redress this situation. The next section looks at the literature on voluntary approaches to environmental protection and the analytic framework used in assessing the effectiveness of voluntary agreements. Details of German industry's agreement to limit CO_2 emissions and its place within the context of overall German climate change policy follows. Attention then turns to the implementation of the voluntary agreement in the power sector.

Voluntary agreements and environmental protection

Conventional wisdom holds that environmental regulation has adverse effects on international competitiveness, economic growth, and employment as production costs associated with environmental regulations reduce competitiveness in global markets. Rigorous environmental standards push industry abroad (industrial flight) and developing countries attempting to compete provide pollution havens. Though a number of studies attempt to test these hypotheses have found little evidence to support claims of industrial flight or pollution havens (OECD, 1993; Blazejczik, Kohlhaas, Seidel, Trabold-Nübler, Löbbe, Walter, & Wenke, 1992), concerns about the impact of environmental regulations on competitiveness persist. Industry continues to oppose strong regulatory measures; governments hesitate to press ahead in the face of such resistance. One response has been the increasing prominence of voluntary agreements in the policy mix of many countries.

The turn to voluntary agreements reflects (in large part) growing disillusionment with so-called command-and-control regulation. While proponents of regulatory standards have emphasized such advantages as effectiveness in achieving environmental goals, public participation and transparency in the decision-making process, and accountability on the part of the administrators, critics argue that regulatory standards are economically inefficient and environmentally ineffective: the costs of pollution control are neglected; the adversarial and legalistic nature of traditional regulatory strategies slows the formulation process and often impedes effective implementation and enforcement; command-and-control regulation hampers the type of innovation that encourages pollution prevention rather than end-of-pipe approaches (Harrison, 1999; Rehbinder, 1997). In contrast, voluntary agreements offer a potential antidote to such shortcomings.

Among the advantages attributed to voluntary agreements are the ability to reduce the overall costs of environmental protection, stimulate innovation, and encourage prompt implementation. With voluntary agreements, firms have greater flexibility in the method and timing of activities, thus facilitating greater efficiency as well as stimulating innovation. Moreover, voluntary approaches further understanding and trust between government and industry, thereby promoting faster implementation and compliance (IEA, 1997; Wicke, 1997; Rehbinder, 1997; EEA, 1997; EU, 1996). However, there are certain caveats and reservations attached to this policy instrument.

The ability to stimulate innovation is likely to correlate closely with the stringency of the agreement but, according to critics, the likelihood that voluntary agreements will move beyond what industry would have done anyway is quite remote. For one, given the dynamics of the negotiating process, environmental goals inevitably will be watered down, that is, they will fall short of what could have been achieved with formal laws or ordinances. Further, the nonbinding and unenforceable nature of most voluntary agreements, combined with the threat of free-riders, leaves the effectiveness of voluntary agreements in question. Finally, beyond the issue of effectiveness, normative concerns have been expressed about a process that allows industry to take over the public policy process, raising questions about democratic control, public access, and transparency (Rennings, Brockmann, & Bergmann, 1997; IEA, 1997; Torvanger & Skodvin, 1999).

Clearly, the debate over the effectiveness of voluntary agreements has been joined but studies that shed greater light on this issue are quite limited. This may be due to the relatively recent introduction of voluntary agreements in many countries. However, another critical factor is the fundamental inattention devoted to program evaluation (Harrison, 1999). In other words, any attempt to sort out conflicting claims about the environmental effectiveness of voluntary agreements requires some attention be paid to the criteria one applies to this question.

Implicit in the critique of voluntary agreements is that performance ought to be measured against what environmental goals could have been achieved through the use of traditional regulatory instruments (or alternatives such as green taxes). Proponents of voluntary approaches often prefer to have effectiveness judged on the basis of improvements over some base year. Neither approach provides the best method of evaluation. The former often assumes optimal results from administrative regulations or economic instruments, leaving aside questions about political viability and/or the gap between idealized as opposed to actual rigor in implementation (see Rehbinder, 1997). The latter may simply reflect developments (e.g., market fluctuations, economic cycles, structural changes, and so forth) that have nothing to do with the voluntary agreement itself. In other words, a critical issue is how causal links between voluntary agreements and environmental outcomes are established.

One method is the use of historical trends that provide a baseline against which to evaluate the results of a voluntary agreement (Storey, 1997); if benefits exceed what can be anticipated based on historical trends, the improvements are ascribed to the voluntary agreement. Conversely, if current performance fails to measure up to past trends, then the voluntary agreement would be judged ineffective. This approach is not without its shortcomings, since the conditions facing industry may be dramatically different than in the past, making such extrapolations problematic. As a consequence, a more fruitful approach to establish this cause-effect relationship is to compare performance against a business-as-usual baseline that indicates what would have happened absent any voluntary agreement. This is the primary method used to assess the environmental effectiveness of the voluntary agreements in this chapter. There are, however, conditions common to the majority of voluntary agreements that complicate this approach: voluntary agreements have often lacked clear targets or deadlines; many make no provision for reporting or monitoring; participants themselves attribute actions to voluntary agreements that would have been undertaken in any case; or they exaggerate claims of effectiveness since, for the vast majority, their primary motive for joining was to avoid regulation (EEA, 1997; Harrison, 1999; Storey, 1997). In other words, the design of voluntary agreements may be a critical factor in determining whether voluntary agreements are effective. Among the elements most commonly cited in this context are:

- consensual rather than adversarial policy styles
- quantified targets
- clearly established monitoring and reporting procedures

- independent assessment of outcomes
- a credible threat by the state to use stronger measures if parties fail to adhere to agreements
- the ability within industrial associations to sanction its members (an ability that is enhanced where membership is homogenous and concentrated)
- contractual provisions that are legally binding on both parties. (EEA, 1997; EU, 1996; IEA, 1997; Kohlhaas & Praetorius, 1996; Wicke, 1997)

All told, certain design features enhance prospects for environmentally effective voluntary agreements. A consensual policy style in which a climate of mutual trust exists between public and private sectors, facilitates the negotiation of more environmentally effective voluntary agreements. At the same time, mutual trust and the carrot of foregone regulatory measures are not enough to ensure compliance; greater transparency and public scrutiny through the adoption of clear targets and reliable monitoring, reporting and verification procedures improve these chances. A credible stick in the form of additional state regulatory measures provides further encouragement to parties tempted to defect or cheat; binding contractual arrangements backed by legal sanctions serve the same purpose. Finally, when industrial associations are able to sanction its members, threats posed by the free-rider problem are much reduced.

In light of the foregoing, the following questions will inform the analysis of German industry's voluntary commitment to reduce its CO_2 emissions:

- What were the most important factors that led to the agreements between the federal government and industry (e.g., the availability of such carrots as holding in abeyance further regulatory measures; a climate of mutual trust between industry and government)?
- Once established, are there credible sticks behind the door available to government in the absence of compliance by industry?
- Are there provisions for monitoring and assessment? How effective are they?
- Are voluntary agreements contractually binding?
- Have industrial associations attempted to mitigate the potential for free-rider problems arising from noncompliance of its members? If yes, what have they done and how successful have they been? If not, why not?

- Do the commitments agreed to, in fact, push industry to efforts that move beyond business as usual or do they simply represent the outcome of trends in energy consumption or actions taken for reasons having little to do with CO_2 abatement? What specific measures have been implemented to help reach the target?

Before undertaking a detailed treatment of German industry's declaration on global warming, however, it is useful to place it within the broader context of Germany's domestic, regional and international commitments to mitigate climate change.

International climate change agreements and domestic policy

Throughout the series of international negotiations leading to the United Nations Framework Convention on Climate Change (UNFCCC) and the Kyoto Protocol, Germany has been a strong proponent of ambitious targets. During preparations for the first session of the Conference of the Parties (COP-1) in Berlin, for example, Germany pushed for a substantive new commitment on reductions in greenhouse gas (GHG) emissions. In advance of COP-3, the EU called for a 15 percent cut in CO_2 emissions by 2010; the U.S. position entering the negotiations at Kyoto was a reduction in emissions back to 1990 levels between 2008–2010. This EU target was based on a burden-sharing arrangement in which Germany was a major contributor—a 25 percent reduction in domestic CO_2 emissions, which translated into an estimated 80 percent of total EU reductions. In the aftermath of the compromise reached at Kyoto, the burden-sharing arrangements negotiated within the EU called for Germany to cut domestic emissions 21 percent—albeit from the basket of six GHGs stipulated in the Kyoto Protocol rather than from CO_2 alone. In other words, these international obligations are closely connected to domestic policy—policy shaped by commitments to dramatic reductions in CO_2 emissions.

As early as 1990, the German government committed to reduce CO_2 emissions 25–30 percent by 2005 (base year of 1987), one of the most demanding targets among industrialized states. Attention subsequently turned to questions about how the agreed target should be achieved. In such questions, fundamental differences arising from the nature of the climate change issue itself had to be resolved: no single government ministry could control global warming policy, each ministry had different organizational responsibilities and constituencies, and with coalition governments the norm in Germany, ministers frequently had different party affiliations.

The two major protagonists in governmental efforts to formulate a policy toward global warming were the Federal Ministry for the Environment (BMU) and the Ministry for Economics (BMWi), which is responsible for energy policy. For the most part, debate focused on the application of a CO_2/energy tax and its linkage to an EU-wide climate protection tax. The BMU favored the adoption of a tax or levy, even in the absence of agreement at the EU level. The BMWi, for its part, opposed a CO_2 levy, especially if it were undertaken unilaterally. In the end, the position of the BMWi won out. In the absence of an agreement among the member states on an EU CO_2/energy tax, a national tax was off the table for the time being.

Despite this setback for those wishing quick action on global warming, other measures were adopted by the government to reduce CO_2 emissions, but they failed to represent the kind of commitment required if Germany were to meet its the reduction goal of 25–30 percent by 2005.[1] Though CO_2 emissions had declined by 14.7 percent between 1987 and 1993 (BMU, 1994a, p. 10), this was due largely to the effects of unification: inefficient energy use in the former East Germany—combined with its reliance on lignite (70–80 percent of primary energy)—meant the shift to other fuels and their more efficient use reduced CO_2 emissions substantially; the collapse of the East German economy led to lower CO_2 emissions as well. According to studies commissioned by the government, however, the measures approved and studied to date fell well short of the 2005 target (BMU, 1994b, p. 87).

This became more problematic for a German government whose commitments to CO_2 reduction came under increasing domestic and international scrutiny as host to COP-1 in 1995. As a consequence, the German government sought to introduce additional measures to bolster the credibility of its global warming policy.

German industry's declaration on the prevention of global warming

Throughout much of the early debate over a CO_2/energy tax, influential voices within industry pushed a very hard line. The powerful Federation of German Industry (BDI)—the umbrella organization for big industry—at first opposed any fee or tax proposal. It was joined by the other major industrial organization, the German Chamber of Industry and Commerce, as well as individual associations from the energy sector and energy-intensive industries. Once it became clear the federal government was moving toward the adoption of some kind of levy or tax, however, they altered their tactics. Rather than die-hard opposition that threatened to marginalize their influence in future efforts to flesh out the government position, emphasis was placed on the necessity of adopting a CO_2 levy/tax only within an international framework: EC-wide agreement was not enough to avoid competitive disadvantages in the world market; agreement must be at the OECD level. (e.g., BDI, 1991; VDEW, 1991).

In a statement issued in November 1991 the BDI emphasizes the global nature of the climate change problem and calls for an approach that would not adversely impact German industry's international competitiveness—an approach creating a favorable investment climate that encourages the development of new and efficient technical processes. For the BDI, this meant firms should not be saddled with CO_2/energy taxes or regulatory standards that diverted scarce resources away from such investments (BDI, 1992). In other words, industry wanted the government to give priority to voluntary measures.

The BMU, BMWi, and representatives from industry entered into exploratory discussions about a voluntary agreement but progress was slow as circumstances changed. The deepening economic recession and a developing debate about Germany's attractiveness as a business location (see Deutscher Bundestag, 1993)—combined with declining public attention to the issue following the Earth Summit—weakened the hand of the BMU within the government as well as in its negotiations with industry. With the prospect of a CO_2/energy tax and other regulatory measures pushed by the BMU much diminished, efforts to have industry put forward a more detailed proposal for a voluntary agreement were rebuffed, essentially terminating this initial attempt to get a voluntary agreement by the end of 1993. The issue was soon revisited, however, as Germany prepared to host COP-1 in the spring of 1995.

This high profile event, preceded by a winter punctuated by two one hundred year floods in Germany, gave the global warming question greater prominence on the domestic political agenda. The German government, as host, also felt compelled to make a credible demonstration of its ability to meet its reduction targets. In the months leading up to COP-1, the government renewed its efforts to negotiate a voluntary agreement with industry. Industry also appeared more interested in a negotiated agreement as the government's threat to wield its taxation and regulatory powers grew more credible. Whether or not an agreement could be successfully negotiated, however, remained an open question, given recent experience and institutional setting.

For the most part, Germany approximates the policy style believed to facilitate the negotiation of effective voluntary agreements—a consensual policy process in which industry is well represented (see, e.g., Hancock, 1989; Katzenstein, 1985). Moreover, one of the guiding principles of German environmental policy is cooperation, in which "the Federal Government will . . . involve industry at an early stage in the opinion forming and decision making process within environmental policy (Federal Ministry for the Environment: 20–22). Historically, there has been a good working relationship between industry and the Economics Ministry and perhaps because of this type of relationship, the BMWi is favorably disposed to the use of voluntary agreements. Relations between environmental policymakers and industry, on the other hand, have been more problematic (see Wallace, 1995). Whereas the BMWi and industry have common interests (e.g., furthering economic growth), the BMU's overriding organizational goal is protection of the environment. For the BMU, the traditional method to achieve this goal is the use of administrative regulations that frequently targets the polluting practices of industry. This, perhaps inevitably, led to a more adversarial relationship, one often characterized by mutual mistrust; a legacy that had to be overcome if the negotiations were to succeed. Another legacy was distrust of voluntary agreements among many officials within the BMU itself.

Against this institutional background, negotiations between industry and representatives from the BMU and BMWi began in January 1995. Representing the BMU in these negotiations was an official who had developed a good relationship with industry officials through his work in the Interministerial Working Group on CO_2 Reduction. Industry was represented by officials from the Federation of German Industry, who worked with separate industrial associations. They in turn represented the individual firms belonging to their associations. The BMWi assumed a somewhat moderating role at times during the negotiations. Among the major points of contention were the explicitness of the commitments, how demanding they should be, and what concessions government would provide in return.

While industry's opposition to a CO_2/energy tax has been well documented, of even greater concern was a proposed Heat Utilization Ordinance for industrial companies. If adopted, it would have required companies to recover and utilize heat generated in their plants and make surplus heat available to others, an expensive process that—in the eyes of German industry—would severely compromise its competitiveness. Industry wanted these measures off the table. The government, for its part, wanted high absolute targets that represented reductions that moved well beyond business as usual. Industry representatives initially offered vague figures because it proved difficult to get concrete numbers from the various associations. When more explicit figures were produced, industry favored specific rather than absolute targets—that is, reductions calculated on a per unit of output basis (specific) rather than in lower overall emissions (absolute). Industry also wanted a commitment establishing 1987 as the base year, since they wanted to include all reductions resulting from unification. The government, on the other hand, was preparing to announce at COP-1 a revision of its national target—25 percent CO_2 reductions by 2005, using 1990 rather than 1987 as the base year. This, at once, would bring it into conformity with the base year employed in the ongoing international negotiations surrounding the UNFCCC and make it more ambitious, since many of the wall fall benefits from unification would be lost.

On March 10, 1995, the "Declaration by German Industry and Trade on Global Warming Prevention" was issued.[2] In this declaration, fifteen industry associations agreed to use special efforts to reduce specific CO_2 emissions or specific energy consumption up to 20 percent by the year 2005 (base year of 1987). For its part, the government agreed to hold in abeyance additional regulatory measures (such as the Heat Utilization Ordinance) and CO_2/energy tax. While welcomed by many concerned about the absence of action on global warming, the agreement was not without critics. Studies from the Wuppertal Institute for Climate, Environment, Energie and the German Institute for Economic Research (Fischedick, et al., 1995; Kohlhaas, Praetorius, & Zeising, 1995), for example, were critical of several elements of the agreement:

- in foregoing the introduction of the CO_2/energy tax and heat utilization ordinance, government was giving up general policy instruments that would affect the entire economy for an agreement that covered a relatively few energy-intensive industrial sectors;
- targets were not clearly defined (e.g., reductions in CO_2 emissions or energy consumption up to 20 percent);
- the commitments to reduce emissions or energy consumption in specific rather than absolute terms could actually result in higher overall emissions if the level of production increased;
- the goal was not demanding enough since the reductions called for were generally lower than longer-term trends in energy consumption underway since the 1970s (i.e., no special efforts would be required to meet such targets);
- there were no clearly articulated means available to the state to sanction those who do not adhere to the agreement;
- the commitments of the industrial associations did not contain specific obligations for individual members and, therefore, would be difficult to enforce;

- no objective monitoring mechanism was established to verify compliance.

In response to such criticism, further negotiations between government and industry resulted one year later in what was portrayed by the Environment Minister as a considerably improved agreement. Included in the revised agreement was a pledge to reduce specific CO_2 emissions by 20 percent, with a change in the base year from 1987 to 1990. Moreover, some of the associations switched their commitments from specific to absolute emissions reductions. Also, additional industrial associations joined, meaning that approximately 80 percent of German industry's total energy consumption was now covered by the agreement. Finally, a monitoring system, to be administered by an independent third party (the Rhine-Westphalia Institute for Economic Research), was established to provide greater transparency in evaluating compliance.

Though addressing several of the issues raised by earlier critiques, the revised agreement was viewed by some as falling short in several critical areas (see Kristof, Ramesohl, & Schmutzler, 1997; Rennings, Brockmann, & Bergmann, 1997; Storey, 1997): no special efforts will be required to meet targets that are still not defined clearly enough; the ability of associations to sanction their members remains weak, thereby contributing to the free-rider problem; there is no credible threat on the part of the state in instances of noncompliance, and it is not easy to detect noncompliance since an early warning system is not built into the agreement.

Whether the shortcomings asserted here—or those cited in our general discussion of voluntary agreements earlier—have diminished the environmental effectiveness of this type of policy instrument will be examined through a study of the German electricity sector's commitment to reduce its CO_2 emissions.

The declaration of the VDEW on climate protection

Among the fifteen (and later nineteen) industrial associations that were parties to the voluntary agreements negotiated with the German government was the Association of German Electricity Supply Companies (VDEW). The power generation sector and the sectoral agreement it concluded with the federal government merit particular attention for several reasons. Most significantly, power production from public utilities is the single largest source of national greenhouse gases, contributing around one-third of Germany's total CO_2 emissions. Furthermore, in contrast to most of the other industry declarations, the VDEW committed to absolute—rather than specific—reductions in CO_2 emissions.

Finally, in the revised agreement of 1996, the base year and target years were shifted to conform to those of the German government's. In the initial 1995 declaration, the formulation was up to 25 percent by 2015 relative to 1987; this subsequently was amended to 12 percent (90 million tons [mt]) by 2015 relative to 1990 and 8 to 10 percent (23–29 mt) by 2005.

This willingness to enter into a voluntary agreement with the government reflected lessons learned from the industry's approach to the galvanizing environmental issue of the early 1980s, forest die-back. In the face of growing public pressure on government to adopt measures that would reduce emissions causing acid rain, the VDEW and its members energetically opposed any actions—to no avail, as it turned out, since the government imposed strict regulations on emissions from power plants anyway. Given this experience, a strategy that engaged the government in a dialogue about the contributions the power sector could make in achieving Germany's global warming objectives was viewed as being much more productive. At the same time, however, the VDEW made clear it expected certain concessions in exchange for its cooperation. Accordingly, a number of conditions were attached to the targets contained in the agreement, which revisited several long-standing concerns of the electricity producers: the political future of nuclear power in Germany; the imposition of regulatory measures such as the Heat Utilization Ordinance; and the use of an energy/CO_2 tax. Specifically, the VDEW stipulated these targets were contingent on several factors:

- that a social consensus is reached on the use of nuclear power on the basis of prevailing law;
- that currently operating nuclear plants be allowed to extend their operating life and increase their output;
- that the completed nuclear power plant at Mülheim-Kärlich be allowed to begin operation;
- that present power plants be guaranteed uninterrupted operation;
- that utilities are free to choose the fuels used in their power plants; and
- that the scope of entrepreneurial involvement in influencing consumer demand is not curtailed by regulation (VDEW, 1997, p. 3).

In other words, the VDEW asserted it would not be able to achieve the targets contained in the agreement unless these conditions were fulfilled.

If these conditions were not met, the reduction target would be lowered to between five and eleven million tons of CO_2.

The VDEW arrived at these targets by circulating a questionnaire to the largest members (around ten) who together constituted the vast majority of generating capacity within the association. In this questionnaire, each utility indicated what measures it was undertaking, in the years up to 2005, that would lead to CO_2 reductions, how many power plants were to be closed and/or opened over the coming years, and so forth. A working group within VDEW was created to determine what overall targets were possible, given information supplied by its members. In this regard, however, it is important to keep in mind that only the VDEW was party to the agreement; nowhere in the VDEW Declaration is there a breakdown of which utilities are to make what contributions to the overall reduction targets. In other words, individual utilities within the VDEW have no reduction quotas or formal obligations to cut CO_2 emissions.

General conclusions drawn from the survey indicated demand for electricity would be relatively flat over the next several decades, meaning that hardly any power plant construction was anticipated until 2015, the year many plants were expected to reach the end of their operating life. This notwithstanding, the primary areas singled out by the VDEW for CO_2 reductions were the replacement of old coal-fueled power plants with newer, more efficient coal-burning plants and the substitution of coal by natural gas. Other areas mentioned were optimizing the use of existing nuclear power plants and bringing on line the Mülheim-Kärlich nuclear reactor, increasing the efficiency of existing power plants, more renewable energy, and greater use of cogeneration and demand-side management (DSM). Of these various sources for CO_2 reductions, the replacement of old with new power plants had by far the greatest potential according to VDEW documents (VDEW, 1995; Grawe & Hoffmann, 1997). Given the different types of fuel mixes characterizing the largest utilities in Germany, this meant that the vast majority of reductions in CO_2 emissions were to come from a select few:

- Bayernwerk and Hamburgische Elektrizitäts-Werke, the fifth and eight largest utilities in Germany, relied almost totally on nuclear power to generate their electricity, meaning that they were marginal to the agreement.
- The energy mix for Energie Baden-Württemberg (fourth largest)—with nuclear power and hydro-power supplying around 76 percent of its electricity—provided little room for absolute CO_2 reductions as well.

- From the perspective of Berliner Kraft und Licht (seventh largest), its primary means to reduce CO_2 emissions were the expansion of district heating and the construction of a natural gas-fueled power plant, measures that were undertaken before the VDEW agreement.
- Vereinigte Elektrizitätswerke Westfalen (sixth largest), generated around 61 percent of its electricity from coal-fueled power plants and 27 percent from nuclear plants. Its ability to replace older coal-fueled plants with more efficient newer ones—or substitute coal with natural gas—were viewed as quite limited, however. An extensive program to retrofit its coal-fueled plants was undertaken in the 1980s to meet the emissions standards adopted in response to concerns about acid rain; this was accompanied by a general overhaul of the plants. To cover the cost of these investments, company managers estimated that the plants must operate until at least 2013.
- PreussenElektra (second largest) was the largest nuclear power producer in Germany with 60 percent of its electricity coming from this source. Hard coal represented the next highest share of power generation, but the limitations on new power plant construction were similar to those described above. Given its large coastal area, PreussenElektra had approximately 70 percent of the wind-generated electricity in Germany, but this was still a minuscule share of overall power generation.
- Vereinigte Energiewerke or VEAG (third largest), the utility that supplied the eastern part of Germany, generated around 93 percent of its electricity from lignite-fueled power plants.[3] The collapse of East German industry following unification, along with substantial increases in energy efficiency produced by the introduction of market forces, resulted in a precipitous decline of over 40 percent in the demand for electricity. In addition, all power plants were required to meet the rigorous federal emissions standards by 1996, which meant they had to be retrofitted at great expense. These factors resulted in the decision to shut down most of the smaller power plants rather than have them retrofitted and to build several new state-of-the-art lignite-burning power plants to cover the remaining production gaps. The closure of the small, inefficient power plants, combined with the construction of these efficient new plants, accounted for a substantial share of the VDEW's CO_2 reduction target.
- RWE Energie, the largest utility in Germany, was part of a powerful industrial group with interests in mining, petroleum and

> chemicals, and engineering and construction. Over 52 percent of its electricity was generated from lignite supplied by Rheinbraun, a mining company belonging to the RWE group. Couched in terms of a CO_2 reduction program, an agreement between RWE/Rheinbraun and the state government of North Rhine-Westphalia was signed in 1994 to open up a new lignite mine to replace a Rheinbraun mine that was approaching exhaustion. In return, RWE committed to undertake measures to increase the operating efficiency of twenty-one operating lignite power plants and replace older, less efficient ones with technologically advanced lignite-burning power plants RWE had been developing. Around one-third of the CO_2 reductions committed to by the VDEW derived from this agreement alone. Another substantial part came from the assumption the Mülheim-Kärlich nuclear reactor owned by RWE would be allowed to start up.

In sum, while the VDEW does not provide a breakdown in the relative contributions of individual utilities in achieving the targets of the voluntary agreement, it is apparent the vast majority of the CO_2 emissions reductions were to come from measures undertaken or planned by VEAG and RWE. This meant it was fairly easy for the VDEW to assess whether its members were implementing measures that would allow the association to meet its target, though no mechanisms existed to sanction members not doing what they said they would. It also implied that concerns about free-riders disrupting the agreement were minimal since most of the utilities were not expected to make much of a contribution in achieving the targets and those that did achieve the targets did so for reasons other than the voluntary agreement. On the other hand, this raised questions about the degree to which the commitment to undertake special efforts to reduce CO_2 emissions was fulfilled[4]—an issue at the heart of the review process agreed to in the 1996 revision of the voluntary agreement.

An independent third party, the Rhine-Westphalia Institute for Economic Research (RWI), was chosen to administer the monitoring system that would evaluate compliance. Since the agreement never clearly defined the criteria used in determining what constituted special efforts, this task fell to RWI as it prepared the first monitoring report. The approach RWI adopted distinguished between actions on the basis of whether they were necessarily brought about by other legal instruments, could have been induced by increases in energy prices, or had already been initiated before the declaration of German industry on climate protection in March 1995 (Hillebrand, Buttermann, & Oberheitmann, 1997, p. 6). All actions falling into one of these categories would have been implemented without the voluntary obligation and cannot therefore be counted among the special efforts.

The second major conceptual task that needed to be addressed in the assessment process was how to establish the business-as-usual baseline. To ensure comparability over time, RWI determined that factors such as the influence of energy prices on energy consumption and investment behavior, the impact of weather (warmer or colder than normal temperatures) on energy consumption, and rates of economic growth needed to be accounted for.

Adjusting for these various factors, RWI found that the public electricity supply sector achieved just under 70 percent of its target as early as 1995 and 80 percent in 1996. Also, in conjunction with the aforementioned criteria, RWI concluded these reductions were not only the result of actions identified by the VDEW, but were in part "the result of legal requirements or effects of the restructuring process in East Germany" (Hillebrand, Buttermann, & Oberheitmann, 1997, p. 40). In the second monitoring report, RWI determined the public power sector had reduced CO_2 emissions 10.1 percent (Buttermann, Hillebrand, & Lehr, 1999, p. 150). Similar to the conclusions of its first report about power plant closings in East Germany, RWI maintained the closings in West Germany cited by the VDEW were due more to the normal replacement cycle for aging plants than to the voluntary agreement (Buttermann, et al., 1999, pp. 154–55). Given the VDEW, as well as several other industry associations party to the voluntary agreement, had almost achieved or exceeded their goals (for the VDEW, 8–10 percent reductions by the year 2005), the report concluded such results were not unproblematic for the persuasiveness of the voluntary agreements "unless this were the motivation to define new and more ambitious goals" (Buttermann, Hillebrand, & Lehr, 1999, p. 157).

The VDEW took exception to some figures included in RWI's calculations. Association representatives, moreover, argued that it was dangerous to evaluate each year, inferring reductions one year translated into further reductions out to 2005. Rather, one had to use a longer-term perspective which accounted for fluctuations. In other words, only in 2005 will it be possible to properly evaluate the rigor of the target (Personal Communication, June 29, 1998 and July 23, 1999). These views of industry notwithstanding, the clear implication of the monitoring reports was that more must be done if the agreement was to move beyond business as usual. Aside from these assessments, critics of the agreement have questioned the credibility of the government's 25 percent reduction target if the sector responsible for approximately one-third of all CO_2 emissions in Germany only reduces from 8–10 percent (Rennings, Brokmann, & Bergmann, 1997, p. 18).

The government consistently took the position that "negotiation on further development of the voluntary commitment will take place each time the annual report is presented" (BMU, 1998, p. 33). In the wake of national elections in September 1998, however, there were questions about the commitment of the new SPD/Green coalition government to voluntary agreements, since a number of representatives from both parties had been critical of the agreement while in opposition. Following the elections, the coalition agreement negotiated between the SPD and the Greens included a statement supporting the use of voluntary agreements, but several other sections of the agreement represented potential sticking points in efforts to negotiate more ambitious emissions reduction targets, most importantly a commitment to phase out nuclear power and a call for ecological tax reform.

Following several months of negotiations among coalition partners and various stakeholders, an eco-tax went into effect on April 1, 1999, raising the price on gasoline, heating oil, natural gas, and electricity; the consumption of coal, however, was exempted (for details, see chapter 6 in this volume). Since eco-tax reform was to be revenue-neutral, the government also announced cuts in company payroll contributions (e.g., statutory pensions). As a consequence, industry representatives did not claim these energy taxes abrogated the voluntary agreement, although they have consistently argued it compromises the competitiveness of electricity generated in Germany.

The debate over nuclear power has a long history in Germany (see, e.g., Hatch, 1986, 1991, 1996). Earlier failed attempts to achieve an energy policy consensus, a broadbased political understanding of nuclear power that went beyond party lines, were resuscitated by the new SPD/Green coalition government following the 1998 elections in efforts to negotiate the conditions of a phaseout, which were long and difficult. A final agreement was announced in June 2000, but it took another year before the Chancellor and the power company executives put their signature to it.

The agreement established the maximum amount of power each nuclear reactor was allowed to generate over its lifetime. Generally speaking, this translated into thirty-two years of operating life, meaning all reactors would be closed by around 2021. The operating life of some plants may extend beyond this time, however, due to a provision that allows operators to shut down less efficient plants ahead of schedule and transfer the remaining production capacity to more efficient ones. The government provided assurances of undisturbed operation within this framework and the power industry agreed not to seek monetary damages in court, even in the case of the Mülheim-Kärlich plant, which has never been allowed to begin commercial operation. In light of the prominence nuclear power played in the VDEW's Declaration on Climate Protection, however, the phaseout agreement was likely to have some effect on the position staked out by industry in negotiating revised voluntary commitments.

In March 2000 the Council of Environmental Advisors announced that the government would not be able to achieve its emission reduction goals unless additional efforts were undertaken, a fact subsequently acknowledged in statements by both the Ministers of Economics and the Environment. In October 2000, one month before negotiations on the Kyoto Protocol were to resume at COP-6 in the Hague, the government announced a further iteration in Germany's Climate Protection Program (BMU, 2000). In order to achieve its 25 percent reduction target, the government estimated an additional fifty to seventy million tons of CO_2 emissions had to be eliminated by 2005; the power industry's share was set at twenty to twenty-five million tons. Among the catalogue of actions in the Climate Protection Program that would affect the power sector most directly were:

- a law adopted in March 2000 that was designed to protect existing cogeneration facilities through 2004 by setting a minimum price of nine pfennig per kWh above market for electricity from cogeneration plants;

- a proposed law that would increase the amount of electricity produced by combined heat and power facilities with the help of a quota system requiring utilities to supply a minimum amount of electricity from cogeneration plants (initially set at 12 percent of total purchases, but up to 24 percent by 2010);

- a renewable energy law passed in April 2000 designed to increase the share of electricity generated from renewables (50 percent by 2050) by setting a minimum price for energy from renewable sources; and

- a voluntary agreement that called on industry to make further reductions in CO_2 emissions.

Negotiations on revisions to the voluntary agreement began in earnest during the first months of 2000. In November of that year a general agreement was signed between the government, BDI, and individual industrial associations that committed industry to specific CO_2 reductions of 28 percent by 2005 (the earlier agreement had set the target at 20 percent) and a 35 percent reduction in emissions of Kyoto gases (expressed in CO_2 equivalents) by 2012 compared to what they were in 1990. It was estimated that this would result in an additional ten million ton reduction in CO_2 emissions by 2005 and a further 10 million tons CO_2 equivalent by 2012 (BMU, 2000). Though the VDEW signed this agreement, negotiations on the specifics of the power sector's contribution to the agreed targets had reached an impasse. From the perspective of industry, the main points of contention were various initiatives pushed by the SPD/Green government since taking office that made CO_2 emissions reductions more difficult and put domestic energy producers at a competitive disadvantage as German and European electricity markets were opening up: the phaseout of nuclear power, ecological tax reform, the renewable energy law, and the legislation promoting cogeneration.

The Third Monitoring Report was released later in November 2000. Covering the years 1997 and 1998, it concluded that the power sector's goal of an 8–10 percent reduction in CO_2 emissions by 2005 had been exceeded by 120 percent in 1997 and more than 124 percent in 1998 (Buttermann & Hillebrand, 2000, pp. 188–191). Echoing the conclusions of the second report, the Third Monitoring Report found that "These results are not unproblematic for the persuasiveness of the voluntary commitments unless they serve as a stimulus to undertake new and more ambitious goals." (Buttermann and Hillebrand, 2000, p. 193)

Negotiations with individual industrial associations were supposed to have been concluded by the end of 2000, at the latest. In the case of the VDEW, however, they continued into 2001. Though representatives from the power sector raised concerns about the constraints the nuclear phaseout imposed on efforts to reduce CO_2 emissions as well as the impact of the eco-tax and the renewable energy law on energy prices and competitiveness, the primary bone of contention (and greatest stumbling block)in these negotiations were the cogeneration laws.

From an ecological perspective, cogeneration provides an attractive option to reduce CO_2 emissions in the power sector. Waste heat from electricity generation can be applied to residential heating or for use in office buildings, hospitals, and so forth, pushing efficiency rates up to 90 percent compared to the 35–38 percent generally provided by conventional power plants. Added benefits are realized when cogeneration plants use natural gas, the least CO_2-intensive fossil fuel. But aside from the environmental attractiveness of this technology, political forces were at play as well.

With the decline in prices following the liberalization of the electricity sector in 1998 (more on this later), many of the cogeneration facilities, owned primarily by municipal utilities, were being priced out of the market. There were increasing pressures on government officials from the industrial association representing municipally owned utilities

(VKU) to help preserve this form of power generation—pressures resulting in the law passed in March 2000 to protect current electricity production through cogeneration. The major utilities, on the other hand, were opposed to such intervention in the power sector. They were especially concerned about the proposed legislation to establish mandatory quotas that would raise the current share of electricity from cogeneration (approximately 12 percent) to around 24 percent by 2010. The primary concern of industry was that mandatory quotas would necessitate the closure of existing power plants—many whose costs had already been written off—to make room for the mandated increase in production from cogeneration (in a market already suffering from surplus capacity), and that market shares would be lost to municipal utilities. In January 2001, the major utilities proposed the alternative Action Program-Climate Protection.

The program put forward by the power generators was a set of voluntary measures that they claimed would reduce CO_2 emissions up to forty-five million tons by 2010, if no further interventions in the market were undertaken. Though criticized by environmental groups and the Greens, it provided a basis for further negotiations. These negotiations were complex, involving intraindustry (primarily the VDEW and VKU), intragovernmental (BMWi and BMU) as well as government/industry bargaining. The Minister of Economics was initially more sympathetic than the Environmental Minister and his colleagues from the parliamentary Green Party, who favored mandatory quotas. The bottom line for the Greens, however, was that a voluntary agreement on CO_2 emission reductions from cogeneration was only acceptable if it explicitly included a provision for a regulatory approach containing mandatory quotas if an agreement was not adhered to.

A compromise was ultimately worked out among the various parties based on a model designed to encourage the modernization and construction of cogeneration facilities by subsidizing the price of electricity from cogeneration. The industry proposal was submitted in May 2001; acceptance by the government was announced in June. Formal approval came with the adoption of implementing legislation by the Bundestag in January 2002.

The final agreement called for a twenty-three million ton reduction in CO_2 emissions (base year 1998) by 2010, but specified that CO_2 emissions must be reduced a minimum of twenty million tons by 2010 through the maintenance, modernization, and new construction of cogeneration facilities. Construction of new facilities would depend on the outcome of a monitoring process, though the agreement does contain a technology-forcing element pushed by the BMU—the provision of subsidies for such small, decentralized technologies as fuel cells.

An interim reduction target of ten million tons was established for 2005. Implementation of the agreement was to be monitored by an independent third party in conjunction with the monitoring of the November 2000 voluntary agreement. If an evaluation undertaken at the end of 2004 raised questions about whether the 2005 target would be reached, the government would initiate regulatory measures by January 2006 to achieve the agreed goals, the preferred method being mandatory quotas. On the other hand, if the agreement is being implemented successfully, the government is obligated not to introduce further regulatory measures that impact the options available to industry. Along the same lines, the implementing legislation for the agreement will replace the Cogeneration Law adopted in March 2000.

In addition to the twenty million ton reduction in CO_2 emissions from cogeneration, the industrial associations committed themselves to a further 25 million ton reduction in CO_2 emissions by 2010, in accordance with the target of forty-five million tons established for the energy sector in the national Climate Protection Program of October 2000. Reflecting the concerns of industry about the nuclear phaseout, possible increases in CO_2 emissions due to the nuclear power agreement would not be included in this commitment to reduce emissions by twenty-five million tons.[5]

Following announcement of the agreement, the two ministers most actively involved in the negotiations characterized it as a satisfactory solution; it also received praise from members of the Greens and SPD in the Bundestag. Representatives from the power sector, on the other hand, indicated they had agreed to the compromise only with great pain and that it would require the greatest of sacrifices from industry. Whether the disproportionate satisfaction implied in these statements reflects the degree to which special efforts will be required to meet the agreed targets rather than mere public profiling for their respective constituencies remains to be seen. It does appear, however, that the primary concerns of the various parties to the negotiations were met. The government received commitments on CO_2 reductions that moves it closer to achieving the 2005 target of 25 percent as well as help meet its share of EU reductions under the Kyoto Protocol. Also included was the potential threat of mandatory reductions through expanded cogeneration capacity should industry fail to follow through on its commitments. For their part, power generators avoided the imposition of mandatory quotas for electricity produced by cogeneration plants, termination of the earlier Cogeneration Law, a commitment from government not to impose further regulations, and an acknowledgement that industry's obligations to reduce CO_2 emissions does not included potential increases from the nuclear phaseout agreement.

Conclusion

The voluntary agreements between the German government and the electrical power generation industry have become a centerpiece of Germany's climate change policy. In many respects, the agreement with the electric utility industry is a model for others in that it contains most features necessary for effective action:

- The relationship between industry and government, generally speaking, has been characterized by a consensual policy style. While historically less the case in the relationship between the BMU and industry, the constraints imposed by economic recession in recent years has made it more open to such collaborative approaches.
- The VDEW's Declaration on Climate Protection and its revisions in 1996 and 2000/2001 contained quantified targets; monitoring and reporting procedures were established and an independent assessment system was implemented. In their interaction, each element serves to enhance the transparency of the agreement. This, in turn, has pushed industry and government alike to address questions related to environmental effectiveness.
- In terms of the so-called stick behind the door, the threat of the Heat Utilization Ordinance or, more importantly (in the case of the power generation sector), a CO_2/energy tax induced industry to return to the negotiating table in the months prior to COP-1. After the agreement, the CDU/FDP government consistently maintained it would resort to direct regulatory action if industry failed to live up to its end of the bargain. Since the government was out of office within two years of the revised declaration there was never much of an occasion to test the credibility of this threat. However, given that certain elements within both parties of the SPD/Green governing coalition viewed voluntary approaches with some skepticism prior to taking office, an argument could be made that such a threat became more credible with the change in government. In negotiations leading up to the 2000/2001 revisions, passage of the March 2000 Cogeneration Law and the announced intention to legislate the expanded use of cogeneration through mandatory quotas has put the utilities under considerable pressure to accept more ambitious targets. The provision in the new agreement for resort to mandatory quotas in the absence of industry compliance in-

dicates the importance attached by the government to sticks as well as carrots in voluntary agreements.

- Despite the absence of controls within the industrial association that could induce compliance of its members, free-riding was not viewed as much of a problem. The primary factor has been the concentration of efforts within two utilities who are undertaking measures for reasons that had little to do with the voluntary agreement. One additional factor, according to some observers, was the relative modesty of the earlier reduction targets.
- Similarly, the fact initial voluntary agreements were not contractually binding on the partners did not seem to have been very significant, given the VDEW had already achieved its reduction targets well in advance of 2005. On the other hand, the more ambitious agreements concluded in 2000/2001 appear to connote some type of contractual relationship signed by both industry and government rather than taking the more conventional form of a self-declaration by industry.

Despite the presence of most features important for the effectiveness of voluntary agreements, questions remain about whether the agreements will be sufficiently effective to produce the kind CO_2 emissions reductions necessary to achieve its national target. Part of this uncertainty relates to developments in the European and international arenas that have significant implications for the power generation sector. The other part has to do with the most recent agreements and their implementation.

Shortly after the initial voluntary agreement had been concluded between government and industry, a revolution in the power generation sector was set in motion with a 1996 directive from the European Union requiring member states to open up their electricity markets. (Until then, the power sector in Germany was governed by a 1935 law that allowed utilities to operate as regional monopolies.) This directive was translated into national law when a new law became operative in April 1998. Rather than being monopoly suppliers to closed markets, German utilities now had to compete with each other as well as with foreign companies for customers.[6] This has certain implications for CO_2 emissions reductions. Companies worried about costs are likely to shift investments away from capital-intensive investments with long amortization periods—i.e., efficient technologies with short lead times (e.g., combined cycle gas turbines that produce 50 percent less CO_2 than comparable coal plants) will be preferred over such capital-intensive options as coal and nuclear. Investment in renewables with high capital costs may be adversely affected and lower prices from liberalization may have the same effect. On the other hand, easier market access may make cogeneration by industry more attractive; investment in certain types of demand-side-management projects may also be attractive since they would not entail high up-front capital expenditures. In other words, the effects of market reforms on efforts to reduce CO_2 emissions are very complex, given the different type of fuel mixes present in individual utilities and the competitive pressures now buffeting them from all sides.[7] One thing is apparent, however. For industry officials, any actions to limit greenhouse gas emissions are now viewed first and foremost through the lens of a liberalized energy market and how those actions may impact competitiveness.

With the adoption of the Kyoto Protocol in December 1997, questions have been raised about how such flexible mechanisms as tradable permits, joint implementation, and the clean development mechanism might be employed in conjunction with the voluntary agreement. Implicit in the use of these instruments is some kind of accounting system that would credit utilities for reductions achieved through such activities. Will this open the door to some kind of arrangement allocating quotas to individual utilities?[8] If so, this could have significant implications for the future of the voluntary agreement.

Finally, whether the recently adopted voluntary agreements can and will deliver the promised reductions in CO_2 emissions remains an open question. If the targets are as ambitious as portrayed by industry, it may increase the potential for free-riding which, in turn, could undermine the commitment of other utilities to adhere to the agreement. Within this context, the fact that voluntary agreements have not been contractually binding may have greater significance than in the past. The forces unleashed by liberalization of the electricity market, when combined with more ambitious targets and the absence of legally binding commitments, have the potential to alter calculations made within corporate headquarters.

To conclude, it is premature to make a final judgment about the environmental effectiveness of the agreement between the German government and electricity producers. Much depends on whether or not it represents a static document written in stone or a dynamic process that allows ongoing corrections and modifications. In this respect, the negotiations leading to the announcement of new voluntary agreements in 2000/2001 are instructive. The uncertainties introduced by a deregulated electricity market, initiatives to phase out nuclear power, and the possible use of the flexible mechanisms proposed in the Kyoto Protocol complicated an already complex set of conditions. Reconciling the interests of industry, various elements within government and their constituencies has been—and continues to be—a difficult task. Nonetheless, the outcome of the most recent negotiations indicates a process capable of movement towards more ambitious goals. It may be prudent, however, to first see whether the promises made in the most recent voluntary agreements can be translated into concrete achievements before making a more definitive judgment about the usefulness of this policy instrument in achieving environmental objectives.

Notes

1. For a more detailed treatment of global warming policy in Germany, see Hatch (1995).

2. In Germany, voluntary agreements typically have taken the form of unilateral declarations or self-commitments on the part of industry though they are generally the result of intensive negotiations with the appropriate ministries.

3. VEAG was created from the power generation sector of the former German Democratic Republic (GDR). Following unification, it was decided that all the nuclear power plants in eastern Germany were to be closed and work terminated on those under construction, since they were not up to federal safety standards. There were also questions about whether or not the lignite industry—the primary source of energy in the GDR—should be kept alive. Ultimately, it was decided not to shut down the entire industry, but rather concentrate production in certain regions. To guarantee an outlet for this production, VEAG was required to burn the lignite in its power plants. In return, VEAG was allowed to operate in a protected market through the year 2003.

4. As indicated earlier, one of the potential sources for CO_2 reduction cited by the VDEW was demand-side management (DSM). Given their monopoly status, large utilities in Germany historically have had a pronounced supply-side mentality, with little faith in end-use efficiency improvements. (Collier, 1994). Accordingly, significant improvements in this area might be one indication of special efforts. The VDEW has said the number of DSM projects increased dramatically in the mid-1990s (431 by 1996, 200 percent more than in 1994). The association, however, was unable to quantify the effects of these measures in term of CO_2 reductions, perhaps reflecting tokenism still accorded to end-use measures by the large utilities.

5. According to estimates contained in the national Climate Protection Program, approximately 8 billion kWh would have to be replaced by 2005, 19 billion kWh from 2006–2010, and 87 billion kWh from 2011–2020. These figures equate to around 3–7, 7–17, and 33–74 million tons, respectively, of CO_2 emissions, depending on the energy source used to replace nuclear power (BMU, 2000).

6. An exception was made for VEAG because of the potential for stranded investment costs related to the decision to construct new lignite-fueled power plants and modernize others. Since this decision was due largely to political forces that wanted to preserve jobs in eastern Germany, VEAG was allowed to keep its market closed until 2003.

7. Reflecting the turbulence unleashed by deregulation, a series of mergers reshaped the electricity sector. Veba and Viag—owners of PreussenElektra and Bayernwerk—merged to form E.ON, which supplanted RWE as the largest utility in Germany until RWE took over Vereinigte Elektrizitätswerke Westfalen. RWE had also sought to acquire a 25 percent share in Energie Baden-Württemberg, but was outbid by the state-owned utility Electricité de France, which was attempting to gain a presence in the German market. Most recently, the Hamburgische Elektricitätswerke led an effort that will merge it with VEAG and Berliner Kraft und Licht to form Germany's third largest utility by 2003.

8. Reflecting such concerns, the directive for an EU emissions trading scheme agreed to in December 2002 permits certain sectors and companies to apply to opt out of emissions trading until 2008 and allows companies to pool their emissions rights.

References

Blazejczak, J., Löbbe, K., Kohlhaas, M., Seidel, B., Trabold-Nübler, H., Walter, J., & Wenke, M. (1992). *Umweltschutz und Industriestandort: Der Einfluss umweltbezogener Standortfactoren auf Investitionsentscheidungen.* Umweltbundesamt Bericht/93, Berlin: Erich Schmidt Verlag.

Bundesministerium für Umwelt, Naturschutz und Reaktorsicherheit (BMU). (1998, February). Decision of the federal government of 6 November 1997 on the Climate Protection Program of the Federal Republic of Germany, on the basis of the Fourth Report of the CO_2 Reduction Interministerial Working Group (CO_2 Reduction IWG). Bonn.

Bundesministerium für Umwelt, Naturschutz und Reaktorsicherheit (BMU). (1990). *Beschluss der Bundesregierung vom 7. November 1990 zur Reduzierung der CO_2-Emissionen der Bundesrepublik Deutschland bis zum Jahr 2005.* Bonn.

Bundesministerium für Umwelt, Naturschutz und Reaktorsicherheit (BMU). (1994a, September). *Environmental policy: Climate protection in Germany, first report of the government of the Federal Republic of Germany pursuant to the United Nations Framework Convention on Climate Change.*

Bundesministerium für Umwelt, Naturschutz und Reaktorsicherheit (BMU). (1994b, November). *Environmental policy: The federal government's decision of 29 September 1994 on reducing emissions of CO_2, and emissions of other greenhouse gases in Federal Republic of Germany.*

Bundesministerium für Umwelt, Naturschutz und Reaktorsicherheit (BMU). (2000). *Nationales Klimaschutzprogramm: Beschluss der Bundesregierung vom 18. Oktober 2000.* Fünfter Bericht der Interministeriellen Arbeitsgruppe "CO_2-Reduktion."

Bundesverband der Deutschen Industrie (BDI). (1992). Initiative of German business for world-wide precautionary action to protect the climate. In *Inter-National Environmental Policy—Perspectives 2000.* May.

Bundesverband der Deutschen Industrie (BDI). (1991). Stellungnahme zum Fragenkatalog der Anhörung der Interministeriellen Arbeitsgruppe 'CO_2-Reduktion' der Bundesregierung zum CO_2-Minderungsprogramm aufgrund des Beschlusses der Bundesregierung vom 7. November 1990. Köln, 2 May.

Buttermann, H. G., Hillebrand, B., & Lehr, L. (1999). *CO_2-Emissionen und wirtschaftliche Entwicklung: Monitoring-Bericht 1998.* Untersuchungen des Rheinisch-Westfälischen Instituts für Wirtschaftsforschung, Heft 28, Essen.

Buttermann, H. G., & Hillebrand, B. (2000). *Klimaschutzerklärung der deutschen Industrie unter neuen Rahmenbedingungen: Monitoring-Bericht 1999.* Untersuchungen des Rheinisch-Westfälischen Instituts für Wirtschaftsforschung, Heft 37, Essen.

Collier, U. (1994). *Energy and Environment in the European Union.* Avebury Studies in Green Research, Aldershot.

Deutscher Bundestag. (1993, September). Bericht der Bundesregierung zur Zukunftssicherung des Standorts Deutschland, Drucksache 12/3620.

European Environment Agency (EEA). (1997). *Environmental Agreements: Environmental Effectiveness. Environmental Issues Series.* Copenhagen 3(1).

European Union (EU). (1996, November 27). *Communication from the Commission to the Council and the European Parliament on Environmental Agreements.* COM(96) 561 final.

Federal Ministry for the Environment, Nature Conservation, and Reactor Safety. (no date). *Umweltpolitik: Guidelines on anticipatory environmental protection.* Bonn.

Fischedick, M., Kristof, K., Ramesohl, S., & Thomas, S. (1995, July). Erklärung der deutschen Wirtschaft: Königsweg oder Mogelpackung? Wuppertal Institut für Klima, Umwelt, Energie, 39.

Grawe, J., & Hoffmann, T. (1997). Stand der Umsetzung der CO_2-Selbstverpflichtungserklärung der deutschen Stromversorger. *VGB Kraftwerkstechnik,* 77(3), 161–165.

Hancock, M. D. (1989). *West Germany: The Politics of Democratic Corporatism.* Chatham, NJ: Chatham House Publishers.

Harrison, K. (1999). Talking with the donkey: Cooperative approaches to environmental protection. *Journal of Industrial Ecology, 12*(3), 51–72.

Hatch, M. T. (1991). Corporatism, pluralism and post-industrial politics in West Germany. *West European Politics, 14*(1), 73–97.

Hatch, M. T. (1996). Nuclear power and postindustrial politics in the West. In J. Byrne & S. M. Hoffman (Eds.), *Governing the atom: The politics of risk.* New Brunswick, NJ: Transaction Publishers.

Hatch, M. T. (1986). *Politics and nuclear power: Energy policy in Western Europe.* Lexington: University Press of Kentucky.

Hatch, M. T. (1995). The politics of global warming in Germany. *Environmental Politics, 4*(3), 415–440.

Hillebrand, B., Buttermann, H. G., & Oberheitmann, A. (1997). *CO_2-Monitoring der deutschen Industrie—ökologische und ökonomische Verifikation.* Band 1: Ergebnisse und Bewertung, Untersuchungen des Rheinisch-Westfälischen Instituts für Wirtschaftsforschung, Heft 23, Essen.

Hillebrand, B., Buttermann, H. G., & Oberheitmann, A. (1997, November). *First monitoring report: CO_2-emissions in German industry 1995–1996.* RWI-Papiere, Nr. 50, Essen.

International Energy Agency (IEA). (1997). *Voluntary actions for energy-related CO_2 abatement.* Paris: OECD/IEA.

Katzenstein, P. J. (1985). *Small states in world markets: Industrial policy in Europe.* Ithaca, NY: Cornell University Press.

Kohlhaas, M., & Praetorius, B. (1996, November). Economic aspects of voluntary agreements for CO_2-emission reduction. Deutsches Institut für Wirtschaftsforschung.

Kohlhaas, M., Praetorius, B., & Ziesing, H-J. (1995). German industry's voluntary commitment to reduce CO_2 emissions—No substitute for an active policy against climate change. *Economic Bulletin, 32*(5).

Kristof, K., Ramesohl, S., & Schmutzler, T. (1997, March). Aktualisierte Erklärung der Deutschen Wirtschaft zur Klimavorsorge: Grosse Worte, Keine Taten? Wuppertal Papers, 71.

Organization for Economic Cooperation and Development (OECD). (1993, January 28–29). Summary report of the workshop on environmental policies and industrial competitiveness. Paris: Author.

Personal communication, June 29, 1998 and July 23, 1999.

Rehbinder, E. (1997). Environmental Agreements: A new instrument of environmental policy. European University Institute, Jean Monnet Chair, Paper RSC No. 97/45.

Rennings, K., Brockmann, K. L., & Bergmann, H. (1997, March). *Voluntary agreements in environmental protection-experiences in Germany and future perspectives.* ZEW, Discussion Paper No. 97-04E.

Storey, M. (1997). *Voluntary agreements with industry.* Annex I Expert Group in the UNFCCC, Working Paper No. 8. Paris: OECD.

Torvanger, A., & Skodvin, T. (1999). Implementing the Kyoto Protocol: The role of environmental agreements. CICERO Report 1999: 4, 30 April.

Vereinigung Deutscher Elektrizitätswerke (VDEW). (1995, November 23). CO_2-Minderungsmöglichkeiten der deutschen Stromversorger. *VDEW Arguments.*

Vereinigung Deutscher Elektrizitätswerke (VDEW). (1997). Report on the "Declaration of VDEW on Climate Protection." A-08/97.

Vereinigung Deutscher Elektrizitätswerke (VDEW). (1991, May 21). VDEW-e.V. nimmt Stellung zur Einfürung von CO_2-Abgaben. Frankfurt am Main.

Wallace, D. (1995). *Environmental policy and industrial innovation: Strategies in Europe, the USA and Japan.* London: Earthscan.

Wicke, L. (1997, January). Umweltbezogene Selbstverpflichtungen der Wirtschaft—Chancen und Grenzen für Umwelt (mittelständische), Wirtschft, und Umweltpolitik. Arbeitspapier Nr. 4. Institut für UmweltManagement.

CHAPTER SIX

Ecological Tax Reform in Germany

Economic and Political Analysis of an Evolving Policy

MICHAEL KOHLHAAS AND BETTINA MEYER

Introduction

On April 1, 1999 a first step towards an ecological tax reform (ETR) became law in Germany. This law was the starting point for one of the projects to which the German government, in power since October 1998, had attached a great deal of importance. In December 1999, the German parliament passed a law on the Continuation of the ecological tax reform that defines another four steps of the ecological tax reform which came into force at the beginning of the years 2000 to 2003. In December 2002, another law modified some elements of the tax reform as of January 2003.

Some leading members of the two coalition parties called it a central project of the modern age and expected it to spur a technological, ecological and social modernization of society that eases the burden on the environment as well as economic and social problems. Curiously, however, both opponents and proponents of the general concept have strongly criticized the laws. That includes business associations as well as labor unions, economists, environmentalist, and politicians ranging from the opposition parties to even some members of the coalition parties.

This chapter analyzes the reform project mainly from an economic point of view. Many economists denounce it as inefficient. This judgment, however, neglects the legal and institutional restrictions that in some cases prevented a first-best solution. The chapter analyzes at which points and why measures depart from the idea that economists have developed of an ideal ETR and identifies the scope for improvements. It will first sketch the general concept of an ETR . It will then outline the main features of the reform in Germany and analyze the most controversial issues of the discussion in Germany. The chapter concludes with a discussion of the perspectives on an ETR.

The concept of an ecological tax reform

The need to strive for sustainable economic development, which preserves the basis for human existence, is now widely recognized. Natural resources must be used more economically and the burden on the environment eased. The traditional instruments of environmental policy—in particular, the use of government regulations, which predominates in Germany—are inadequate for this purpose and induce economic costs that are unnecessarily high. Alternatively, an ecological reform of the taxation system involving the greater use of environmental charges could help to achieve environmental objectives more efficiently and foster technological innovation for environmentally friendly products and production processes.

The idea of environmental charges or taxes goes back to Pigou (1920) who recognized that the market mechanism fails to allocate resources efficiently, if external costs are not reflected in market prices. An appropriate tax may help to internalize externalities. Baumol and Oates (1971) showed that taxes can be an efficient tool to pursue environmental goals, even if the externalities cannot be quantified and environmental targets are set politically.

The core concept of an ecological tax reform is the combination of environmental taxes or charges subsequently used to reduce other taxes and charges so that the business sector and private households as a whole will not face a higher overall tax burden (revenue-neutral tax reform). Moreover, substituting the revenue of eco-taxes for traditional taxes or charges can reduce distortions caused by the current tax system. Especially those taxes or charges should be reduced that are considered harmful to the economy. In view of the persistence of mass unemployment, the proposal to reduce social security contributions and thus wage costs has been pursued in Germany. This may offer firms an incentive to raise their demand for labor.

Attaining economic gains such as a reduction of the overall economic cost of the tax system or of unemployment—in addition to an improved environment—has been termed double dividend. A controversial theoretical and empirical debate explored the circumstances under which various forms of a double dividend are possible and likely to occur (see, for example, Bovenberg & de Mooij, 1994; Bovenberg & Goulder, 1996; Arndt, Heins, Hillebrand, Meyer, Pfaffenberger, & Ströbele, 1998; Böhringer, Pahlke, Vähringer, & Voß, 1997; Meyer, Bockermann, Ewerhart, & Lutz, 1999; Parry, Williams, & Goulder, 1999; Bach, Kohlhaas, Meyer, Praetorius, & Welsch, 2002). However, even if there is no double dividend, there may still be good reasons to reduce GHGs and environmental taxes may be an efficient instrument to achieve this. In this context, Goulder, Parry, and Burtraw (1997) showed that even in the absence of a double dividend, taxes may be preferable to other forms of environmental regulation which do not raise revenue.

Key features of the German ecological tax reform

In Germany the two components of the ETR, an increase in taxes on energy and the compensating reduction of social security contributions, have been regulated in separate laws.[1] This means that there is no formal earmarking of the revenue of the eco-taxes. In the law that reduces social security contributions, there is not even an explicit reference to the ETR. There is, however, a strong political commitment that the tax reform be revenue-neutral.[2] Indicating which compensatory changes are to be financed by the revenue of the eco-tax was meant to increase the credibility of the revenue-neutral approach.

Energy taxation

The ecological component of the tax reform in Germany presently consists of a higher taxation of energy products. Energy—or the emissions released through its use—is an important element of virtually all proposals for an ecologically oriented reform of the tax system. A tax on energy is frequently seen as a necessary step towards sustainable development. The current discussion increasingly has come to focus on the threat to the earth's climate posed by energy-related emissions. Yet other environmental damage and risks arising when energy is obtained or used—such as nuclear risk or land use—are also cited.

Many countries are therefore striving to reduce energy consumption or the emissions related to energy use. In 1995, the former government of chancellor Helmut Kohl had set the target of reducing CO_2 emissions by 25 percent up to 2005 compared to 1990. Under the framework of the Kyoto Protocol and the European Union (EU) burden sharing arrangement, the German government accepted an emissions reduction target of 21 percent from a basket of GHG for the period 2008 to 2012. The government of chancellor Gerhard Schröder, however, explicitly endorsed the stricter national target of 25 percent.

An energy tax can help Germany reach these targets but also more demanding goals which are likely to be set in the future. Higher energy prices increase the incentive to use fuels and electricity more efficiently and so to reduce specific energy requirements. Moreover, the tax lowers the economic break-even point for technological and organizational measures to reduce energy consumption, accelerating the pace of energy-saving technological progress. There are, of course, complementary and alternative instruments to pursue environmental targets, such as tradable permits, command-and-control measures, subsidies or voluntary agreements. This chapter, however, will focus only on the potential contribution of eco-taxes and the design of an ETR.

Increase of energy taxes

The ETR in Germany increases taxes on energy in a complex way. On the one hand, they raise existing taxes on petroleum products (gasoline, diesel fuel, heating oil and natural gas); on the other hand, they introduce and increase a tax on electricity. Eco-taxes are levied on final energy consumption. To avoid double taxation, electricity producers receive a rebate for eco-taxes paid on energy sources purchased to produce electricity. It is important to note that in this respect the ETR is different from the taxes on petroleum products that have been levied before and will coexist with the eco-taxes in the future.[3] The former have been levied on inputs to power generation, such as natural gas and heavy fuel oil, too.

The increase in tax rates is 3.07 Euro cents (ct) per liter gasoline and diesel fuel per year, adding up to 15.35 ct within five years. The increase of 2.05 ct per liter heating oil is executed in the first step only. The tax on natural gas was raised by 0.164 ct per kilowatt-hour (kWh) in 1999 and another 0.202 ct at the beginning of 2003. Furthermore, a tax of 1.02 ct per kWh electricity is introduced in the first step and raised by 0.26 ct in the following four steps.[4] Coal will not be taxed.

Provisions for energy-intensive sectors

The government considered it necessary to take steps to ensure energy taxation does not impair the ability of German trade and industry to compete internationally and to avoid economic hardships for certain groups of private households. Therefore, some users are eligible for reduced tax rates:

- Between 1999 and 2002, the goods and materials sector (i.e., manufacturing industry, energy/water, mining, and construction sector) as well as agriculture, forestry, and fishery paid only 20 percent of the regular eco-tax described above (except for motor fuels) if their energy consumption exceeded a certain threshold.[5] Since January 2003 those sectors have to pay 60 percent of the regular tax rates. Mainly private households, retail and private road transport, service companies, public institutions and small enterprises pay the full tax rate.
- Moreover, some enterprises are eligible for tax rebates. Up to 2002, all eco-tax payments of an enterprise in the goods and materials sector which exceeded the savings of social security contributions by more than 20 percent were refunded. From 2003, 95 percent of the additional tax payment in excess of the savings of social security contributions will be reimbursed.
- Eco-taxes on electricity and motor fuels for public transport are only half of the regular rate.
- Up to 2002, electricity for night-storage heaters installed before April 1999 was taxed at half of the regular rate, between 2003 and 2006 at 60 percent of the regular rate. In 2007 the regular rate for electricity will apply.

Renewable energy sources, cogeneration, and small-scale power plants

Special provisions are also made in order to promote environmentally less harmful sources of energy:

- Electricity from renewable sources will be exempt from the tax if the producer uses it himself or if it comes from a network or an electric line that is exclusively fed by renewable sources. The law does not, however, exempt all electricity from renewable sources from the electricity tax. This is mainly due to legal and technical reasons that will be discussed below. Therefore, the government decided to use the tax revenue from renewable energy sources to promote investments in renewable energy sources.
- Combined heat and power plants (CHP or cogeneration plants) with a utilization rate above 70 percent will receive a full rebate of all energy taxes levied on their inputs. This is in contrast to other forms of power generation, which are only eligible for a rebate of

the new energy tax. Thus, the ETR explicitly favors cogeneration compared to other forms of power generation. But the privilege is lower than in the case of a primary energy tax which taxes electricity generation proportionally to the efficiency of a power plant; with an average efficiency of power generation of 38 percent in Germany, the tax on cogeneration would be about half of the average rate. Small power plants (less than 0.7 megawatt (MW) in 1999 and less than 2 MW from 2000 onwards) are not subject to the electricity tax. This provision is motivated mainly by administrative practicability, but at the same time favors decentralized power generation in small cogeneration plants.

Reduction of social security contributions

The additional revenue of energy taxation is to be used for a reduction of social security contributions. Statutory social security contributions for health, pension and unemployment insurance added up to 42.3 percent of gross wages before the reform. The government intended to reduce this sum by at least 2.4 percent below 40 percent in three steps of the ETR within the legislative period. In 2003 the tax is expected to raise 18.8 billion euro. This allows the government to reduce pension insurance rates by 1.7 percent compared to a situation without ETR. The reduction is split equally between employers and employees. It thus reduces the non-wage costs of employers; employees in turn receive higher net wages. Since demographic and other changes would have required an increase in contributions, the actual pension insurance rates were reduced by only 0.8 percent (from 20.3 percent in 1999 to 19.5 percent since 2003) with the help of government grants to the pension insurance.

Controversial issues

The following section will discuss the most controversial issues in the debate over an ETR.[6] It will focus on the debate over the implemented laws, but it will include some issues that have been important in earlier phases of the debate and thus influenced the design of the current tax reform.

Unilateral tax reform

One of the main issues in the debate over an ETR in Germany has been whether the German government should undertake it unilaterally.

Business representatives and members of the political opposition argued that Germany should not undertake an ETR without at least the EU, or even better, all OECD countries. Two main reasons have been put forward in support of this position: the potential negative effects on the German economy and the prevention of carbon leakage.

- In an open economy, in which enterprises have to face foreign competition (in domestic as well as export markets), an ETR may impair the competitiveness of energy-intensive enterprises and drive them out of business if energy costs rise. This may lead to a devaluation of physical and human capital which enterprises could have otherwise used for a longer time. If this happens very fast, more jobs may be destroyed than created. Especially those regions where energy-intensive sectors have a high share in economic production are likely to experience these problems. Moreover, this implies distributive effects that may appear unacceptable: owners of energy-intensive enterprises, workers who lose their jobs or consumers of energy-intensive products may undergo economic losses.
- From an ecological point of view, the relocation of energy-intensive production impairs the effectiveness of unilateral measures. The reduction of domestic emissions of greenhouse gases may be partially offset or even overcompensated by an increase of emissions abroad. This effect has been termed carbon leakage. The theoretical as well as the empirical evidence about carbon leakage is mixed (see IPCC, 1996). There is, however, a broad consensus that carbon leakage is likely to mitigate the effectiveness of unilateral measures, but will not offset it fully.

Undoubtedly, an ETR would be all the more effective, and the effects on competitiveness would be all the smaller, if more countries participate or take equivalent measures. Therefore, the German government promotes a harmonization of energy taxation in the EU. The chances for effective harmonized measures within the EU are slim, however, and even worse beyond this region. Unilateral measures, however, do not mean going it alone. Almost all European countries use or plan to use taxes and charges as an instrument of environmental policy; a growing number of countries refer to the idea of an ecological tax reform and express interest in a shift from traditional to environmental taxes (e.g., Austria, Belgium, Denmark, Italy, and the Netherlands). The more EU members that pursue this course, the better are the chances for a European-wide tax. Moreover, other countries are likely to undertake more stringent climate policies, creating a more level playing field for German companies, even if they do not introduce environmental taxes.

The ratification of the Kyoto Protocol is very likely, with only Russia's signature necessary for it to come into force. The council of European environmental ministers in December 2002 adopted a directive on emissions trading which is supposed to be launched in 2005. This means more scope for environmental measures without impairing competitiveness, but also more alternative instruments to achieve Germany's targets.

Design of the energy tax

TAX BASE

The eco-taxes are levied on final energy consumption in relation to the quantity consumed (e.g., gallon of fuel or kWh of natural gas). Demands have been made to choose a different tax base, preferably CO_2 emissions or—if that is not possible—the consumption of primary energy. A tax on final energy does not provide an incentive to increase the efficiency of refining or power generation. Therefore, a tax on primary energy would be more efficient and preferable from an economic point of view. There are, however, legal constraints precluding a primary energy tax. If secondary energies (such as electricity or petroleum products) remain untaxed domestically, the principle of nondiscrimination within the EU and the World Trade Organization (WTO) requires that imports of these products remain untaxed, too. This would give a strong incentive to use imported rather than domestically produced secondary energy sources, which would render the tax ineffective.[7]

If the objective is primarily to reduce CO_2 emissions, the most efficient tax base would be the CO_2 emissions or the carbon content of energy sources.[8] Compared to a tax on primary energy, this tax base additionally gives an incentive to switch to less emission-intensive sources of energy. However, similar problems result from foreign purchases of energy as just described in the case of a primary energy tax. Primarily electricity is affected by this problem, since emissions occur exclusively during the transformation process instead of during consumption. With a CO_2 tax, electricity would be taxed indirectly via the fossil fuels used for its production. Imported electricity, however, would not be touched by domestic emission taxes. If there are no equivalent taxes abroad, imported secondary energies would gain a competitive edge compared to domestically produced energy. This effect will be all the stronger, the larger the difference between the total emissions that occur along the energy chain and the emissions related to final consumption.

Hence, proposals have been made to tax secondary energy sources in relation to the CO_2 emissions that are caused along the specific energy chain. This would mean to take into account the efficiency of the transformation process of, for example, a power plant and the carbon content of the energy inputs used. Yet here again, trade-related problems may occur. Usually, information about primary energy input or CO_2 emissions along the energy chain will not be available for imported energy sources. Therefore, it is difficult or even impossible to differentiate the tax rates for imported electricity in relation to the primary energy used or CO_2 emissions released along the energy chain. One country, Finland, wanted to avoid this problem by taxing imports of electricity with the average tax rate on domestic electricity, but lost a dispute concerning this issue at the European Court of Justice in 1998.

Even if a CO_2 or primary energy tax cannot be implemented and the tax cannot be differentiated according to the individual production process, it would still be desirable and possible to differentiate a tax per unit of final energy according to the average CO_2 emissions or primary energy input along the energy chain for different sources of energy, such as coal, natural gas, heating oil, or gasoline. Such a tax would give an incentive to switch from more to less emission-intensive energy sources in final consumption, but not to reduce emissions in power generation or refining. There are no legal or administrative impediments to such second-best taxation. The structure of the German tax rates, however, does not reflect such a concept. Table 6.1 shows the tax increase and the total tax rates in 2003. They reflect substantial differences in tax increases and total tax rates across different sources of energy. The tax rate for electricity is indicated with respect to final energy content as well as with respect to (average) primary energy input.

Several points should be noted:

- Motor fuels are subject to very high rates (gasoline about 2.48 \$/gallon, diesel fuel about 1.78 \$/gallon after the fifth step at an exchange rate of 1 euro to the dollar). Two reasons may justify this: motor fuels also serve to finance the traffic infrastructure and should therefore reflect the respective costs. Moreover, traffic causes other important environmental and nonenvironmental externalities, such as noise, emissions, traffic casualties and so on. Some studies estimate that the costs of these externalities run up to 2.5 euro per liter (9.46 \$/gallon) of gasoline. These values, however, have been challenged and are subject to a heated debate.
- Coal was not taxed in Germany before the tax reform and will not be afterwards. This has been strongly criticized, especially

Table 6.1. Energy taxes before and after April 1, 1999

		Tax rate before April '99	Tax increase 1999 to 2003 (Eco-tax) total			Total energy tax in 2003		
Energy source	Unit	Cent per unit	Cent per unit	Euro/GJ	Euro/ton CO_2	Cent per unit	Euro/GJ	Euro/ton CO_2
Coal	kg	0,00	0,00	0,00	0,00	0,00	0,00	0,00
Heavy fuel oil	kg	1,79	0,71	0,18	2,31	2,50	0,64	8,12
Heating oil	liter	4,09	2,05	0,57	7,77	6,13	1,72	23,28
Natural gas	kWh	0,18	0,37	1,02	18,15	0,55	1,53	27,28
Electricity (final energy)[1]	kWh	—	2,05	5,69	—	2,05	5,69	—
Electricity (primary energy)	kWh	—	—	2,16	36,61	—	2,16	36,61
Gasoline, unleaded	liter	50,11	15,34	4,74	65,84	65,45	20,22	280,85
Dielsel fuel	liter	31,70	15,34	4,29	57,93	47,04	13,14	177,63

[1]Average efficiency (38%) and CO_2 emissions (0.56 kg per kWh)

since the CO_2 intensity of coal is very high. It has been argued that the largest share of coal (about 75 percent) is used in power plants and, as a consequence, would be exempted from the tax in any case. About 20 percent is used by energy intensive industries and would thus be taxed at reduced tax rates or even rebated. Only coal used by private households—less than 5 percent of total consumption—would be subject to the regular tax rate. Therefore, exempting coal from the tax would not do much harm. This argument is true if taxes on coal were treated like other eco-taxes. The taxes that existed before (e.g., on natural gas and heating oil), however, are not rebated even if they are used in power generation or by energy-intensive sectors. As long as the old taxes are maintained, coal should be taxed to some extent in power generation as well, in order to avoid discriminating against other, less emission-intensive inputs into power generation. At least the remaining part, which is mainly used in energy-intensive industries, should be taxed just like other energy sources. There is, however, strong political pressure to protect energy-intensive industries from a high tax burden and coal mining from increased adjustment pressures.

- Heavy fuel oil also is taxed at relatively low rates. This gives it a competitive edge relative to other inputs to power generation and heating.
- The new tax on electricity per gigajoule (GJ) final energy is relatively high compared to the tax on heating oil or natural gas. Relative to the primary energy input or average CO_2 emissions, the difference is far less pronounced. It must be noted, however, that electricity is additionally taxed indirectly, since the taxes on natural gas and gas oil, which existed before the ecological tax reform, are levied on inputs to power generation as well.

Altogether, the tax increases as well as the tax levels do not follow a consistent pattern, neither with regard to energy content nor CO_2 emissions. This clearly impairs the efficiency of the tax with respect to avoiding environmental damages.

The level and path of tax rates

Critics have pointed out that the tax rates of the eco-taxes are too low to induce a sufficient reduction of CO_2 emissions. This argument, however, neglects the need for eco-taxes to be phased in or gradually increased, thereby allowing time for individual polluters and the whole economy to adjust to changing prices and scarcities. With this precaution, the capital stock will not be devalued too fast and energy-saving investments can be undertaken within the normal investment cycle. This way, restructuring the economy is achieved at lower costs and less economic or social friction. Therefore, the precise tax rate of the first steps is not a decisive feature. It is far more important that energy consumers use the time to adjust to future energy taxes. Whenever an investment is made (e.g., a car is bought, a house built or a production line constructed) the technology should be used which is most economical, taking into account expected energy prices during the life cycle of the investment. Moreover, suppliers should develop new products and production processes that allow economizing on energy even more in the future. For this to happen, it is necessary that future increases of the tax rate be announced as far in advance as possible in order to set a clear signal for adjustment measures and to help avoid misguided investments.

After introducing the ETR in 1999 by specifying the first step, the German government subsequently defined another four steps of the ETR extending even into the next legislative period. The last step came into force at the beginning of 2003. Due to strong political resistance, the reelected government postponed a decision about further steps to the year 2004 after an evaluation of the first five steps. This interruption of tax increases rather than the size of the single steps should be the major concern. The need for further measures in climate policy is unquestionable. European emissions trading may take the leading role in reducing industrial CO_2 emissions. For other sectors, such as private households and transportation, no measures are in sight which could replace the continuation of the ETR. In any case, new policies should be announced as early as possible in order to allow energy users to undertake appropriate adjustment measures.

Treatment of renewable energy sources

As discussed above, it was not possible to have an emission tax at this stage, but this has consequences for the treatment of renewable energy sources. With an emission tax, renewable energy sources would be subject to relatively low taxes or even remain untaxed. There have been demands, therefore, to exempt electricity produced with renewable energy sources from the electricity tax. This could be justified by the fact that these emit no or at least substantially less greenhouse gases and other pollutants.[9] The competitiveness of renewable energies would thus be improved, even though the cost of most renewable electricity generation—especially photovoltaics—exceeds the market price for electricity far more than the value of a tax exemption and would need further incentives to become viable.

The German Ecological Tax Reform stipulates a very limited exemption. Electricity from renewable sources will be exempt from the tax if it is used by the producer itself or if it is supplied from a network or an electric line that is exclusively fed by renewable sources. This can be explained by the same reasons that prevented an emission tax in the first place: if imported electricity cannot be treated in the same way as domestically produced electricity, it has to profit from the most favorable treatment applied domestically. Therefore, conditions for an exemption of electricity from renewable resources have been set in a way that can be verified with reasonable effort and reliability, even for imported electricity.

This regulation gives only a very limited incentive to produce electricity from renewable energy sources and has been criticized by proponents as well as opponents of the ETR. Since renewable energies have become a symbol for environmentally friendly technologies, taxing them impairs the credibility of the ETR. The German government therefore will use (part of) the revenue from taxes on electricity from renewable energy sources to promote renewable energy. Moreover, the German parliament passed the Renewable Energy Sources Act which stipulates that grid companies have to buy all electricity from renewable energies above market prices. The obligatory payment is 0.51 euro per kWh for photovoltaics, between 0.06 and 0.09 per kWh for wind energy, and between 0.09 and 0.10 euro per kWh for biomass energy, depending on the size and location of the plant.

Provisions for energy-intensive sectors

One of the main issues is the effect of an ETR on the competitiveness of the German economy and especially of energy-intensive sectors. Energy-intensive sectors will be hit harder by energy taxation and profit less from the reduction of social security contributions than other branches. Therefore, even a revenue-neutral tax reform, though it does not place an extra tax burden on the economy as a whole, may impose substantial costs on some, especially energy-intensive, sectors. As a consequence, these sectors will have higher production costs and may lose competitiveness. As discussed above, this is problematic for two reasons. First, it places adjustment costs on the domestic economy, especially if structural change takes place fast and the existing capital stock is devalued in a short time. Second, carbon leakage may impair the ecological effectiveness if emissions are just relocated.

It is important to distinguish between those two reasons. If the main preoccupation concerns the problems of structural change and the distributive effects associated with it, precautions should be taken to manage, not eliminate, structural change. If a country aims at a permanent reduction of global emissions through unilateral measures, it may want to take permanent precautions against carbon leakage. This, however, does not seem to be very relevant. It does not make sense to undertake unilateral measures in the long run if the environmental problem is global, as in the case of the greenhouse effect. Moreover, even within the framework of international agreements such as the Kyoto Protocol, individual countries are obliged to reduce domestic emissions. A reduction of emissions due to relocation of production is treated in the same way as an increase in energy efficiency or fuel switching. Countries are likely to fulfill this obligation in the cheapest way possible, even if this implies carbon leakage. Therefore, managing structural change seems to be the only valid cause for special provisions in the longer run. Most countries that have introduced energy taxation for ecological reasons have made special provisions for energy-intensive sectors.

Various tax concession concepts are conceivable. There are several important issues with regard to their design (see Bach, Kohlhaas, & Seidel, 1997). The most important issues in the German debate were the definition of enterprises or production processes that are eligible for tax reductions and the tax rates that should be paid by the beneficiaries. The more restricted the demarcation of the beneficiaries, the smaller will be the loss of incentive to reduce emissions and loss of tax revenue. A precise identification of those enterprises which are not able to cope with higher energy prices, however, requires detailed data—for example, about the energy consumption of production processes, available technologies or the competitive situation on the relevant markets. The necessary administrative procedures are likely to be very complicated. Moreover, the process is likely to be subject to lobbying efforts that attempt to safeguard rents and to prevent structural change. On the other hand, the wider the group of beneficiaries, the less complicated administrative procedures need to be, but tax revenue and emission reduction will be lower.

There is no single best-practice design for tax allowances within the context of an energy tax. In selecting concrete models, it is necessary to weigh the partly conflicting demands against each other: reducing the adjustment pressure for energy-intensive sectors, ecological effectiveness, economic efficiency, compatibility with market principles, and issues of administrative feasibility. Such weights cannot be derived from scientific principles but must be determined politically, taking into account political priorities and specific national circumstances. For example, in a large economy with numerous energy-intensive companies like Germany, discretionary policies are more difficult to implement than in smaller economies like Denmark. In the short run, politicians tend to prefer a narrow delimitation of beneficiaries because it promises a higher incentive effect and tax revenue. The price to pay for this approach is discretionary interference with market allocation and complicated bureaucratic processes. If sustainable development is to be achieved in a way that is compatible with the market system, and the idea of eco-taxes as a market-oriented instrument is taken seriously, discretionary special provisions need to be kept to a minimum.

The German ETR established a broad and rules-based system: All companies of the goods and materials sector (i.e., manufacturing industry, energy/water, mining and construction sector) as well as agriculture, forestry and fishery have to pay reduced eco-tax rates (except for motor fuels). Up until 2002 the reduced rates amounted to 20 percent of the regular rates. Moreover, the most energy-intensive companies received compensation for all tax payments which exceeded the reduction of pension contributions by more than 20 percent. With the beginning of 2003, this system has been revised: the reduced rates now amount to 60 percent, and 95 percent of the tax payments exceeding the (simple) savings of pension contributions will be refunded. With these changes, the German government intended to spread the burden of the eco-taxes more evenly and improve their ecological effectiveness. It expressed the opinion that four years after the introduction of the ETR, the incentive to use energy more efficiently could be increased without impairing the international competitiveness of energy-intensive industries. It is not clear at all, however, if the described changes achieve these targets. Figure 6.1 shows the tax schedule of an (arbitrary) enterprise for different quantities of energy consumption before and after the revision.

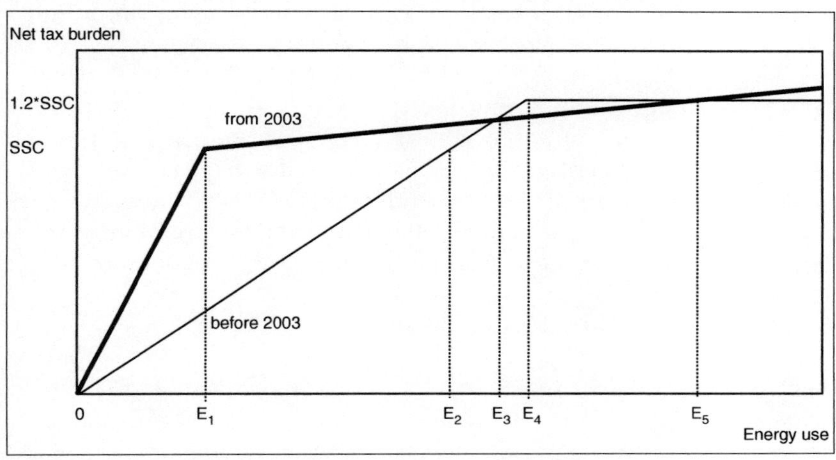

Figure 6.1. Net tax burden before and after 2003

The level of the curve represents the tax burden net of tax rebates, the slopes represent the marginal tax burden (i.e., the tax increase associated with a higher energy consumption). The steeper the curve, the higher the incentive to economize on energy consumption. The overall tax burden is higher in the new tax schedule in all segments except between E_3 and E_5. The slope is steeper between 0 and E_1 (60 percent as compared to 20 percent) as well as above E_4 (3 percent as compared to 0 percent).[10] Between E_1 and E_4 the incentive to economize on energy is lower than previously. Therefore, the net effect is ambiguous. Data about the number of companies in the different segments, their energy use and their sensitivity to price changes would be necessary to estimate the net effect. None of this information was available when the new law was passed in December 2002.

Compensatory reduction of taxes or charges

The use of the additional tax revenue

The additional revenue raised by eco-taxes can be used to increase government spending, to cut other levies, or to reduce public debt. In Germany, there has been a far-reaching consensus among the proponents of an ETR to strive for a revenue-neutral tax reform that leaves the tax burden of business and private households as well as the government budget unchanged. At a period when most countries strive to reduce state intervention and government budgets, it would have been questionable to boost spending via an ETR. Moreover, most proponents agreed that the additional revenue should be used to cut social security contributions in order to cope with high unemployment in the German economy.

This view has been challenged by pointing out that an ETR needs to be accompanied by supporting measures, such as investment in infrastructure (e.g., public transportation) or publicly funded research and development. However, a majority believed that the necessary funds could be raised by restructuring public spending. In particular, cutting ecologically harmful subsidies was supposed to supply the means for supporting measures at the same time as bringing relief to the environment.

Reduction of social security contributions

The reduction of social security contributions compensated for by increased transfers from the public budget has been denounced as a subsidization of the social security system, thus running counter to the principle that all benefits should be covered by contributions. This, however, is not quite true. At present, the statutory social security system is burdened by so-called non-insurance-related benefits. These are benefits not originating from the social insurance program nor, for political reasons, do contributions cover costs according to actuarial calculations—for example, active labor market policies and pensions to retirement age easterners not covered by contributions to the pension funds in east Germany. The German Institute for Economic Research (DIW) estimates that the share of non-insurance-related benefits not covered by government grants were in a range of about 35 to 75 billion euro in 1995 (see Meinhardt & Zwiener, 1997). This corresponded to roughly 10 to 20 percent of the social security contributions.

Noninsurance-related benefits should be financed by general tax receipts. Replacing social security contributions can thus be justified from the point of view of economic theory and should not be considered an undesirable subsidy. It will also include those parts of the population that are exempted from social security contributions—mainly self-employed persons and civil servants —in making payments for public tasks that are of general interest, such as the costs of the German unification. Moreover, the reduction of labor costs may give a positive impulse to employment.

Another criticism of such a tax shift is that, in the long run, a reform of the social security system will be necessary, mainly for demographic reasons. Using the revenue to support the system may take away the pressure for reform and delay necessary steps.

Additionally, the link between eco-taxes and the pension system has been attacked frequently in the public debate triggered in 2000 by rapidly rising world prices for petroleum products, especially car fuels. Before this debate, it appeared to be quite popular to link the revenue of eco-taxes to a reduction of social security contributions. Most people recognized labor costs needed to be reduced in order to create employment. This helped create an acceptance of higher energy taxes as long as they appeared to be moderate. Fuel price increases in 2000, due primarily to rising world energy prices and the devaluation of the European currency, changed this attitude. Irritated tax payers, incited in part by the media and opposition parties, asked what fuel taxes have to do with pension payments. Some demanded that energy taxes be used to promote energy savings and subsidize public transport; otherwise, they could not be considered ecological. This discussion displays an enormous lack of understanding about the functioning of the market mechanism and incentive taxes, as well as the contribution of non-insurance-related benefits to the problems of the pension system. Moreover, it is questionable if tax payers are really willing to accept a higher tax burden and a larger government budget. Nonetheless, the debate points to two problems: the marketing of the ETR and the political earmarking of tax revenues on the one hand, and the demand for supporting measures to facilitate the adjustment to higher energy prices on the other.

WILL ECO-TAXES PROVIDE STABLE REVENUE?

The fear has been expressed that eco-taxes will not provide stable revenue: If they were successful in reducing energy use or emissions, the tax base would shrink continuously. A top representative of a business association phrased it this way: Either the tax is ecologically useless or it will not provide any revenue. Therefore, tax policy should not rely on eco-taxes.

Obviously, the extreme version of the statement is wrong. Eco-taxes do not intend to and will not reduce energy consumption to zero. Therefore, there will always be some tax revenue. Nevertheless, questions about the erosion of revenues over time and the degree to which they will be sufficient to finance important public spending are valid. The short answer to these questions is that, for a long time, there should be no real problem. The concept of an ETR usually assumes that the tax will be phased in (i.e., increased steadily over a long period of time). If energy consumption is to be reduced to a sustainable level, a substantial increase in tax rates will be necessary within the next decades. As long as energy consumption is reduced less than the tax rate is increased, total tax revenue will grow.

Problems may occur if no further reduction of energy use and emissions is necessary or, in the case of emission taxes, if renewable energy sources become competitive on a large scale. This, however, is likely to take several decades. This period will be long enough to profit from the potential benefits of an ETR. Other reforms may be necessary afterwards, but no tax system has to last for an eternity. Reforms of the tax system have been quite frequent in the past and probably will be in the future.

The effects of the ecological tax reform in Germany

In 1998 the Federal Ministry of Finance commissioned a model-based impact analysis of the German environmental fiscal reform, addressing the effects on CO_2 emissions, economic growth, employment, and personal income distribution (Bach, Kohlhaas, Meyer, Praetorius, & Welsch, 2002). The analysis used an econometric input-output model (PANTHA RHEI, see Meyer & Ewerhart, 1998) and a dynamic computable general equilibrium model (LEAN, see Welsch & Hoster, 1995) to analyze the macroeconomic effects.

The analysis shows an initial reduction of CO_2 emissions by 0.5 to 0.8 percent relative to a reference scenario. In the medium term the reduction amounts to 2 to 3 percent (see fig. 6.2).[11] This corresponds to no less than 20 to 25 mt per year. Nevertheless, the environmental fiscal reform alone cannot guarantee that the targets for reduction of CO_2 emissions agreed to either nationally or under the European burden sharing agreement will be achieved.

With respect to Gross Domestic Product, the predicted initial change ranges from –0.2 to +0.2 percent, whereas the effect by 2010 is between -0.4 and zero percent. Imported energy will be replaced by energy saving equipment and insulation measures; this implies a shift in demand from imports to domestic output. At the same time labor is substituted for energy and labor-intensive sectors will be favored compared to energy-intensive industries. Simulations with both models produce an increase in employment, but it is more pronounced in

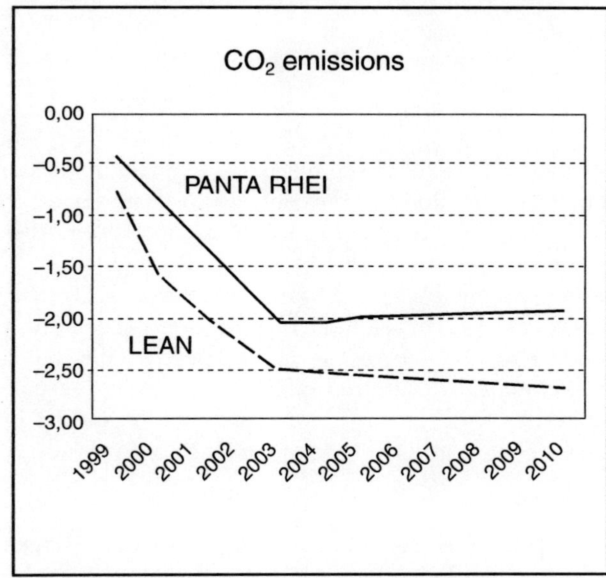

Source: Bach, Kohlhaas, Meyer, Praetorius, & Welsch, 2002

Figure 6.2. CO_2 emissions: simulated effects of the ecological fiscal reform

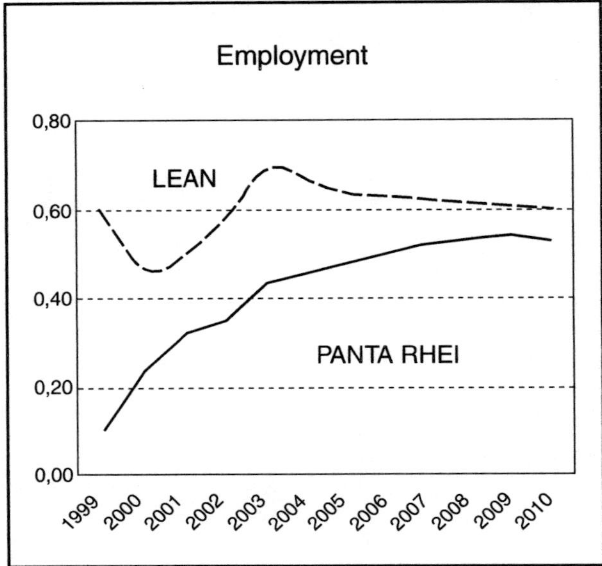

Figure 6.3. Employment: simulated effects of the ecological fiscal reform

LEAN than in PANTA RHEI (see fig. 6.3). The predicted employment effects are between 0.1 and 0.6 percent. In absolute terms, this implies that the environmental fiscal reform could lead to the creation of up to 250,000 additional jobs by 2010.

The ETR in Germany can thus be considered successful in reducing CO_2 emissions and increasing employment, even if the effects are small compared to the size of the challenges by climate change and mass unemployment. The limited effects are mainly due to the careful approach taken by the German government and could be increased if the ETR were to be continued in a systematic and steady way.

Perspectives for the ecological tax reform

The ETR is an innovative approach for Germany, where environmental policy has been dominated by command-and-control measures for decades. The use of environmental levies appeared especially attractive in the field of climate change policy which has to give incentives to a large number of heterogeneous energy users to find innovative ways to maintain welfare with less emissions. The concept of the ETR gained broad support around the mid-1990s, when it became clear that German unification—after an initial positive impulse for employment—would place a heavy burden on the labor market and the economy in general for a long time. Addressing environmental as well as economic and social problems at the same time gained support from those concerned about the environment as well as those who focus on the labor market. As a market-based instrument it appealed to economists and even many conservative politicians who traditionally were reluctant to introduce additional command-and-control measures. Thus, combining two urgent issues helped to receive backing from different camps and made the ETR politically highly efficient.

The German ETR can be considered environmentally effective, even if the induced emission reduction will by itself not suffice to reach the targets of climate change policy. This, however, is in line with both economic research and the position of the German government that a policy mix is necessary and most efficient in this field. Moreover, the environmental effectiveness can be improved by gradually increasing the tax rates. The economic efficiency of the German ETR seems questionable if one looks at the structure of the tax rates and the special provisions, especially for manufacturing. This is true from a first-best perspective, but much more difficult to judge in an imperfect world, where frictions lead to unemployment and international competitors are located in countries which do not contribute to global effort to stop climate change. Taxes certainly are a cost-efficient instrument to give an incentive to the large number of private households which are subject to a uniform tax rate. The response of many producers who offer more energy-saving products since the tax has been introduced (e.g., car manufacturers) indicates that the ETR is likely to spur technological innovation and thus improve the effectiveness of environmental taxation and reduce the costs of emissions reduction in the longer run.

The administrative efficacy of the ETR has been impaired by the special provisions, as well. Nevertheless, the administrative costs are relatively low, particularly since the infrastructure for the administration of taxes on petroleum products had already been in place. Especially in comparison with alternative instruments, notably command-and-control measures, there should be a clear advantage, but no systematic research has been done so far.

Expectations of its advocates had been very high in the beginning, leading to deep disappointment after the law had made its way through the political process. Some of the alleged shortcomings are due to legal, technical, and political constraints. One should therefore be careful when comparing the actual ETR to a first-best solution. Such a comparison may be helpful in guiding the long-term development of the ETR and in setting out an agenda for administrative, political and scientific work. This comparison, however, should not be used to reject a reform project that responds to an urgent need for action in the field of climate protection and employment creation. Rather, one should focus on relaxing the political, legal and economic constraints or finding better ways to achieve the objectives.

Other shortcomings are due to the influence of interest groups. The strongest resistance came from energy intensive industries, especially the chemical industry. The same industries, however, have opposed command-and-control measures in the past and emissions trading in the recent European discussion. Their opposition is directed towards the objectives of environmental policy and their own contribution to them rather than specific instruments. The protest of commuters against fuel prices arose when world market prices increased sharply. This indicates that resistance is triggered by rapid changes (which in this case were not even caused by the ETR) and thus even confirms the concept of gradually phasing in environmental taxes and charges.

In the future, emissions taxes are likely to interact and compete with emissions trading in climate change policies. There are some potential advantages of emissions trading over taxes in the current situation: In particular, there is less cause for concerns about effects on competitiveness within Europe, since emissions trading will be established on the EU level, including the new Eastern European member states. Nevertheless, there are fears that EU member states try to give a competitive advantage to some or all of their industries by allocating generous emissions allowances. Moreover, by giving away emission allowances for free, the burden on energy-intensive industries can be reduced. This, however, implies that no revenue will be created which enables the government to reduce distortions in the economy.

The implementation process of the European emissions trading system shows that this instrument is nonetheless subject to similar lobbying efforts and induced shortcomings as other measures. Furthermore, European emissions trading will not take place in an upstream system that covers all energy users and uses. Therefore, it will not be as efficient as possible and entail far more administrative and transaction costs. This feature creates the need to apply other instruments for energy uses not covered by emissions trading, especially in transportation, private households and small business. For this reason Germany is likely to further develop an ETR. This should be based on a long-term gradual increase of environmental taxes and a more systematic structure of tax rates. In doing so, it will face the challenge to coordinate environmental taxes with complementary and competing instruments of energy and climate policy in an efficient way.

Notes

1. Information on the German ETR in English and other languages is available on the web site of the Federal Environmental Ministry; cf. http://www.bmu.de/files/oekost_en.pdf and especially BMU (2003).

2. The tax reform equals the expected additional tax revenues to the reduction of social security contributions; due to difficulties of estimation, the actual outcome may not be fully revenue-neutral.

3. Eco-taxes on petroleum products have been introduced as an amendment to the law on existing taxes which raises the tax rates. Therefore, it is somewhat arbitrary to call them eco-taxes as opposed to the old taxes. This, however, is helpful, because some rules apply only to the tax increase.

4. On top of energy taxes, the value added tax (VAT) of 16 percent will be raised, so that the total price increase will be larger than indicated by these tax rates.

5. The threshold is a tax payment of 512,50 euro for electricity and 512,50 euro for petroleum products per enterprise and year. Therefore, only 60,000 of 200,000 to 250,000 enterprises of manufacturing industry are likely to profit from reduced tax rates. See Meyer, Bockermann, Ewerhart, & Lutz (1999).

6. This should be considered as a discussion of stylized positions and arguments, not a documentation of the debate.

7. Since Germany has embarked on the liberalization of energy markets, energy users will increasingly be free to profit from price differences for imported energy.

8. It should be noted that a CO_2 tax focuses exclusively on CO_2 emissions. There are, however, more greenhouse gases and other externalities that should be taken into account in environmental policy. Since it is difficult to quantify and monetarize all externalities, an agnostic point of view could be to choose the energy content as tax base.

9. It should be noted renewable energy usually causes some kind of externalities, for example, land use, injury to birds, or sound emitted by wind energy mills. Those externalities, however, are estimated to be substantially lower than those of fossil energy, so renewable energies should still be promoted vis-à-vis fossil fuels.

10. With a rebate of 95 percent, the remaining (marginal) burden of an extra unit of energy will be 5 percent of the reduced rate of 60 percent, i.e., 3 percent of the regular tax rates.

11. This calculation does not take account of the potential reductions through new, highly efficient power-station techniques, which are to be promoted by the environmental fiscal reform.

References

Arndt, H-W., Heins, B., Hillebrand, B., Meyer, E. C., Pfaffenberger, W., & Ströbele, W. (1998). *Ökosteuern auf dem Prüfstand der Nachhaltigkeit.* Berlin: Analytica.

Bach, S., Kohlhaas, M., & Praetorius, B. (1994). Ecological tax reform even if Germany has to go it alone. *Economic Bulletin, 31*(7), 3–10.

Bach, S., Kohlhaas, M., & Seidel, B. (1997). The use of tax allowances to reduce competitive disadvantages resulting from ecological tax reform. *Economic Bulletin, 34*(7), 17–28.

Bach, S., Kohlhaas, M., Meyer, B., Praetorius, B., & Welsch, H. (2002). The effects of environmental fiscal reform in Germany: A simulation study. *Energy Policy, 30*(9), 803–811.

Baumol, W. J., & Oates, W. E. (1971). The use of standards and prices for protection of the environment. *Swedish Journal of Economics, 73.*

Böhringer, C., Pahlke, A., Vähringer, F., & Voß, A. (1997). Volkswirtschaftliche Effekte einer Umstrukturierung des deutschen Steuersystems unter besonderer Berücksichtigung von Umweltsteuern. Institut für Energiewirtschaft und Rationelle Energieanwendung, Band 37, Universität Stuttgart.

Bovenberg, A. L., & de Mooij, R. A. (1994). Environmental levies and distortionary taxation. *American Economic Review, 84,* 1085–1089.

Bovenberg, A. L., & Goulder, L. (1996). Optimal environmental taxation in the presence of other taxes: An applied general equilibrium analysis. *American Economic Review, 86,* 985–1000.

Bundesministerium für Umwelt, Naturschutz und Reaktorsicherheit (BMU). Federal Environmental Ministry. (2003). The ecological tax reform: introduction, continuation and further development to an ecological financial reform. Retrieved January 2003, from http://www.bmu.de/english/download/files/oekost_en.pdf.

Goulder, L. H. (1995). Environmental taxation and the double dividend: A reader's guide. *International Tax and Public Finance, 2,* 157–183.

Goulder, L. H., Parry I. W. H., & Burtraw, D. (1997). Revenue-raising versus other approaches to environmental protection: The critical significance of preexisting tax distortions. *The Rand Journal of Economics, 28*(4), 708–731.

Intergovernmental Panel on Climate Change (IPCC), (1996). Climate change 1995: Impacts, adaptations, and mitigation of climate change: Scientific-technical analyses. In R. T. Watson, M. C. Zinyowera, & R. H. Moss (Eds.), *Contribution of Working Group II to the Second Assessment Report of the Intergovernmental Panel on Climate Change* (Section 11.6.4.1). Cambridge and New York: Author.

Linscheidt, B., & Truger, A. (2000). Ökologische Steuerreform: Ein Plädoyer für die Stärkung der Lenkungsanreize. *Wirtschaftsdienst* 2000/II, 98–106.

Meinhardt, V., & Zwiener, R. (1997). Steuerfinanzierung von versicherungsfremden Leistungen in der Sozialversicherung. *Vierteljahrshefte zur Wirtschaftsforschung, 66* (3–4), 352–361.

Meyer, B., Bockermann, A., Ewerhart, G., & Lutz, C. (1999). *Marktkonforme Umweltpolitik.* Heidelberg.

Meyer, B., & Ewerhart, G. (1998). Multisectoral policy modelling for environmental analysis. In K. Uno & P. Bartelmus (Eds.), *Environmental accountig in theory and practice* (pp. 395–406). Dordrecht, Boston and London: Kluwer Academic Publishers.

Meyer, B., & Kunz, C. (2002). Diskussionspapier zur Steuerbefreiung für Strom aus Erneuerbaren Energien und zur privaten Vermarktung von Ökostrom. Kiel, Februar 2002. Unpublished paper available from the author at bettina.meyer@munL.landsh.de.

Meyer, B. (2002). Ökologisch kontraproduktive Subventionen im Energiebereich. Diskussionspapier/Dokumentation und Hintergrundmaterial zu Vorträgen, Kiel, April 2002. Unpublished paper available from the author at bettina.meyer@munl.landsh.de.

Parry, I. W. H., Williams, R. C., III, & Goulder, L. H. (1999). When can carbon abatement policies increase welfare? The fundamental role of distorted factor markets. *Journal of Environmental Economics and Management, 37*, 52–84.

Pigou, A. C. (1920). *The economics of welfare.* London: Macmillan and Co.

Welsch, H. (1996). Recycling of carbon/energy taxes and the labor market: A general equilibrium analysis for the European community. *Environmental and Resource Economics, 8*, 141–151.

Welsch, H., & Hoster, F. (1995). A general equilibrium analysis of European carbon/energy taxation: Model structure and macroeconomic results. *Zeitschrift für Wirtschafts- und Sozialwissenschaften, 115*, 275–303.

CHAPTER SEVEN

Assessing the Flexible Mechanisms of the Kyoto Protocol

Implications of Joint Implementation and the Clean Development Mechanism for National Policy Instruments

Andreas Oberheitmann

Introduction

Since the United Nations Conference on Environment and Development (UNCED) in Rio de Janeiro in 1992, the international negotiations on project-based mechanisms to mitigate global climate change have evolved from Activities Implemented Jointly (AIJ) in the pilot phase to the establishment of Joint Implementation (JI) and the Clean Development Mechanism (CDM) as flexible instruments of the Kyoto Protocol (Protocol).

Even after the Seventh Conference of the Parties (COP-7) in November 2001, when rules and guidelines were laid out in the Marrakesh Accords, there is still some reluctance on the part of Annex I Parties (industrialized countries) and their private companies to participate in these flexible instruments. This reluctance to ratify the Protocol was due in large part to the uncertain state of climate policy in the United States under the Clinton administration. The Bush administration's declaration not to ratify the Protocol gave a push to the international community to bring the Kyoto Protocol into force. In terms of private sector involvement, the lack of incentives led to lower investment in AIJ projects than originally envisaged. Companies have been reluctant to unconditionally support the flexible instruments due to the high uncertainties involved in the development and assessment of baselines (Michaelowa, 1998a, p. 403). Another factor was the difficulty in formulating an operational definition of additionality.

To counter this reluctance, national governments considered providing incentives in climate policy to promote the participation of private companies. These incentives include the compensation of national environmental tax obligations or offsets of obligations assumed in voluntary agreements by domestic industries to reduce GHG emissions. There are, for example, national programmes currently in the Netherlands, Denmark, and Sweden committing up to $150 million for such purposes (Oberheitmann, 1999b, p. 18).

Generally, any Annex I country can provide incentives for participation in the Kyoto mechanisms—incentives such as no-regrets measures, launching of national programs, or the compensation of national industry obligations with emission credits accruing from JI or the Clean Development Mechanism (CDM). Within this context, however, two main questions arise: (1) What is the relationship between the incentives provided and the Kyoto Protocol's principle of environmental additionality? and (2) Are the possible incentives provided by national governments compatible with the concept of additionality?

In the next pages, these questions will be addressed in detail. Following brief descriptions of JI and the CDM within the context of the international climate change negotiations, the domestic implications that flow from the establishment of incentives are discussed (for example, to what degree taxes can be offset by emission credits from JI and the CDM). The final section discusses the potential international obstacles to the effective operation of JI and the CDM, chief among them being whether no-regrets measures are additional due to the provisions and aims of the Kyoto Protocol.

JI and CDM—Flexible instruments of the Kyoto Protocol

The literature on JI and the CDM is wide-ranging (e.g., Albrecht, 2002). For the purposes of this chapter, however, attention is focused on the provisions in JI and the CDM concerning additionality since this principle represents a potential obstacle to the use of incentives by national governments to attract private company investments in these mechanisms.

JI and the CDM in the context of the international climate change negotiations

The 1992 United Nations Framework Convention on Climate Change (FCCC) established a process that allows further elaboration of rules and mechanisms for international climate policy in the follow-up Conferences of the Parties (COP). Accordingly, at the 1997 COP-3 in Kyoto, the primary aim was agreement on binding reductions or limitations on greenhouse gas (GHG) emissions and to establish suitable policies and measures to fulfil these obligations. The most important result of the Conference was the obligation of the contracting parties listed in Annex B of the Protocol to reduce the greenhouse gases listed in Annex A of the Protocol—carbon dioxide (CO_2), methane (CH_4), nitrous oxide (N_2O), hydrofluorocarbons (HFCs), perflourocarbons (PFCs) and sulphur hexafluoride (SF_6)—in the budget period of 2008 to 2012 by an average of 5.2 percent measured against 1990 levels (1995 for the last three gases listed).

Fulfillment of the obligation was distributed unequally among the contracting parties due to their abilities to reduce greenhouse gases and (possibly) to their negotiating skills. Most Annex B countries committed themselves to an 8 percent reduction, though some countries such as the Russian Federation must only stabilize their emissions; Iceland can emit 10 percent more than 1990 levels. To provide greater flexibility in achieving national GHG emissions reduction obligations, the Kyoto Protocol developed three flexible instruments (FCCC, 1998): (1) Joint Implementation (Art. 6), (2) the Clean Development Mechanism (Art. 12), and (3) Emissions Trading (Art. 17).

Joint Implementation and the Clean Development Mechanism are project-based mechanisms that create emission credits through the reduction of greenhouse gases against a project baseline of a business-as-usual scenario. (For a discussion of emissions trading, see chapter 8 in this volume.)

The Kyoto Protocol comes into force once it is ratified by 55 percent of the Parties and the Annex I countries representing 55 percent of the total Annex I country emissions in 1990 (Art. 25,1). As of January 28, 2003, 104 countries had ratified (approved, accepted, or acceded to) the Protocol, 73 of them are non-Annex I Parties, and 31 Annex I countries (representing about 30 percent of all Annex I countries and 43 percent of total Annex I Party emissions). Without the United States, to reach the Article 25 target, the participation of the Russian Federation, Japan, and Canada is essential. This, however, already led to considerable concessions regarding CO_2-sinks at COP-6bis in Bonn in 2001.

In general, JI means the compensation of a national emissions reduction obligation by comparable emissions reductions in another country. The rationale behind JI is that since it is irrelevant where CO_2 or other GHG are emitted or saved in terms of their possible impact on climate change, it becomes ecologically interesting when there are different marginal GHG reduction costs in investor and host countries. The higher the marginal avoidance cost differences, the more economical GHG reductions can be credited to the account of one's own obligations (Oberheitmann, 1999a, p. 15). JI and CDM are a cost saving alternative to domestic CO_2-emission reduction measures.

According to the PRIMES study (Capross & Mantzos, 2000), the average marginal abatement costs in Europe are 51.2 US$/t CO_2. The reduction costs in JI and CDM projects are in a range of between 0.1 to 19 US$/t CO_2 (see Table 1). However, the relative prices of JI and CDM certificates are different in Europe as domestic marginal abatement costs vary across the European continent from 12.7 US$/t CO_2 in Germany to 142 US$/t CO_2 in the Netherlands.

The object of compensation in JI can vary. It may be a national CO_2-tax payment, the fulfillment of national voluntary agreements, or the purchase of tradable emissions credits. As JI requires a compensation object, it is called an add-on-instrument.

The principle of JI had already been incorporated in the FCCC. Article 3, 3 of the Convention. It states: efforts to address climate change may be carried out cooperatively by interested Parties. Article 4, 2 allows Annex I parties of the Convention to implement such policies and measures jointly with other Parties. The Third Conference of the Parties incorporated this principle into Article 6 of the Kyoto Protocol. Preconditions for the utilization of Joint Implementation are that any such project: (a) has the approval of the Parties involved; (b) provides a reduction in emissions by sources or an enhancement of removals by sinks that is additional to any that would otherwise occur; and (c) does not acquire any emission reduction units if it is not in compliance with its obligations under Articles 5 and 7; and (d) the acquisition of emission reduction units shall be supplemental to domestic actions for the purposes of meeting the quantitative commitments within the Kyoto Protocol.

The resulting Emission Reduction Units (ERUs) compensate for national measures from 2008, the beginning of the first budget period. Any acquired ERUs shall be added to the assigned amount for the acquiring Party and any emissions reduction units that a Party transfers to another Party in accordance with the provisions of Article 6 shall be subtracted from the assigned amount for the transferring Party (Art. 3, 10 and 3, 11). If there is a question about the implementation of the requirements, transfers and acquisitions of emission reduction units may continue, but any such units may not be used by a Party to meet its commitments against the Kyoto Protocol until the issue of compliance is resolved (Art. 6, 4).

Whereas JI (as well as Emissions Trading) can only be used by countries listed in Annex I of the FCCC, the CDM was established for cooperation between Annex I and non-Annex I countries. The rationale and rules of JI and the CDM (Art. 12 of the Kyoto Protocol) are similar; the CDM basically is JI between Annex I and non-Annex I countries. As formulated in the Kyoto Protocol, the aims of the CDM are to assist non-Annex I countries achieve sustainable development, contribute to the mitigation of global climate change, and help Annex I countries reach their emissions reduction targets (Art.12, 2).

Article 12, 8 provides that "the Conference of the Parties serving as the meeting of the Parties to this Protocol shall ensure that a share of the proceeds from certified project activities is used to cover administrative expenses as well as to assist developing country Parties that are particularly vulnerable to the adverse effects of climate change to meet the costs of adaptation." At COP 7 in Marrakesh, concrete rules on CDM were adopted, including a 2 percent CDM fee. According to Article 12, 7, the Conference of the Parties "shall, at its first session elaborate modalities and procedures with the objective of ensuring transparency, efficiency, and accountability through independent auditing and verification of project activities." Certified Emission Reductions (CERs) obtained during the period from the year 2000 up to 2008 can be used to assist in achieving compliance in the first commitment period (Art. 12, 10). Thus, having the same function mode, both JI and the CDM can be used to compensate emissions reduction measures in either Annex I or non-Annex I countries in the same way. For private companies, the crucial difference in the choice of the two mechanisms is in the starting date for the measures (2000 for CDM and 2008 for JI).

The political popularity of flexible mechanisms has changed over time in both Annex I and non-Annex I countries. At the beginning of the Kyoto process, JI and CDM were seen merely as loopholes to evade domestic measures for GHG emission reduction. However, since there is an economic backdrop in many Annex I countries, the potential cost savings of JI and CDM and the existence of concrete rules on the flexible mechanisms have made the instruments more popular for industry and the government. As for non-Annex I countries, there was a similar process. In the past, developing countries such as China feared that being host to CDM projects would reduce the scope for cheap domestic GHG emission reduction potentials should they assume quantitative obligations under the Protocol in the future. Over time, however, the Chinese government has come to favor the CDM as a cost saving means to attract foreign capital and to improve national energy efficiency.

The concept of additionality in JI and CDM

Additionality is crucial for the flexible instruments of the Kyoto Protocol. It creates the unit of the transaction for the utilization of JI and the CDM. Environmental additionality, within this context, means that JI or CDM projects must produce real emissions reductions that would not have occurred in the absence of JI or the CDM. This environmental additionality consists of two parts, the project mandate and project baseline (see Table 7.1).

The project mandate of JI is the reduction of emissions (by sources) or the enhancement of removals by sinks. As CDM projects, sinks are not mentioned explicitly. The reduction of GHG emissions is the only environmental mandate. However, the achievement of sustainable development for developing countries is the second aim of CDM projects (Art. 12, 2). Additionality generates the transaction unit of JI and CDM, creating an additional environmental benefit through fulfillment of the project mandate beyond the project baseline (see Figure 7.1). Over time, autonomous technological progress reduces GHG emission reductions or additional environmental benefits within the project. For this reason, the CDM relevant project duration may be limited.

Article 6, 1(b) provides that Annex I countries can transfer or acquire emission reduction units (ERU) from projects that reduce GHG emissions or enhance anthropogenic removals by GHG sinks, provided the environmental benefits of such projects are additional to any that would otherwise occur (i.e., environmental additionality).

Through projects under the CDM, Annex I Parties can use certified emission reductions from project activities in non-Annex I countries to meet a certain part of their national emissions commitments under the Kyoto Protocol (Art. 12, 3[b]). Similar to provisions for JI, the emis-

Table 7.1. Additionality in JI and CDM

JI:	Any such project provides *a reduction in emissions by sources, or an enhancement of removals by sinks,* that is additionally to *any that would otherwise occur.* (Art. 6, 1[b])
CDM:	*Reductions in emissions* that are additional *to any that would occur in the absence of the certified project activity.* (Art. 12, 5[c])

Source: FCCC/CP/1995/7/Add.1: 19; FCCC/CP/1997/7/Add.1: 13.

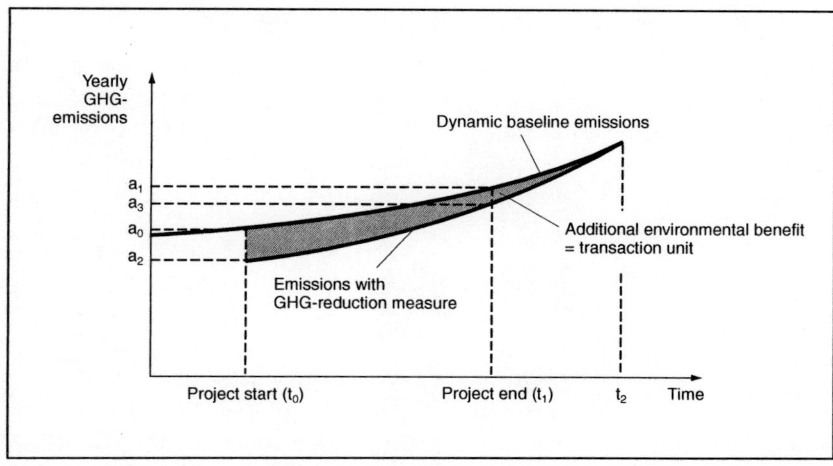

Figure 7.1. Environmental additionality

sions reductions from CDM projects are constituted and certified on the basis of emissions reductions that are additional to any that would occur in the absence of the certified project activity (Art. 12, 5[c]). Enhancing anthropogenic removals by GHG sinks currently not covered by CDM projects. Besides the environmental dimension of additionality, CDM projects are not allowed to substitute official development assistance (financial additionality); safe and sound technology used for the CDM project shall be the best available and practicable for the circumstance of the host party (technological additionality); and the CDM project should not occur in the absence of an emission reduction credit trading system (Project/Program additionality).

Since many non-Annex I countries are concerned that the least-cost potentials for emissions reductions could be lost to Annex I countries via the CDM, there is still some reluctance on the part of these countries to participate in CDM measures. Thus, an important task for future negotiations is to establish the volume or percentage of CERs that remain with the host countries when participating in CDM projects.

Implications for instruments employed in national climate policy

Private companies will only invest in JI and the CDM if certain conditions are met. Governments considering the introduction of incentives for private company participation have to take these conditions into account. Yet, even then, private companies may hesitate to participate in these mechanisms. National governments, as a consequence, may introduce incentives to induce private companies to invest in JI or CDM projects. The introduction of these incentives, however, may have certain implications for national economic policy.

General interest of the investor party and investor party private companies

The add-on quality of JI and the CDM provides incentives for investor governments to participate in the projects. It may be economically rational to maximize the ERUs and CERs since the more additional environmental benefits accruing from JI and CDM projects, the higher the credits to compensate for national actions fulfilling the commitments towards FCCC. There are, however, certain restrictions on the use of the CDM (only a hitherto undefined part of the quantitative national commitments can be compensated through CERs) and JI (the Protocol provides JI measures be supplemental to national actions). This is less restricting than limitations on the CDM, but CDM projects are currently more favorable to investors since crediting is possible from 2000 onwards.

Clearly, investor Party private companies will participate in JI and the CDM only if economically feasible, at least over the long-term. Projects are more attractive if:

- they are already economically feasible, but have not yet been undertaken, due, for example, to uncertainty about the political or economic framework conditions in the host country (no-regrets measures). The Intergovernmental Panel on Climate Change estimates this no-regrets potential at 10–15 percent of global emissions (Michaelowa, 1998b, pp. 10–11). Approval of such measures as JI or CDM projects provides a strong incentive for private companies to participate since, apart from the compensation potential of JI and the CDM, participation in the mechanisms are not connected with economic losses and the JI/CDM status of these projects provides some certainty in terms of being supervised by FCCC bodies.
- they provide cost-efficient compensation for commitments of companies to their national governments, assuming a reasonable degree of certainty. The feasibility of compensation is deter-

mined by the differences in emissions reduction costs, taking the transaction costs into account. The objects of compensation may be domestic energy tax obligations that can be satisfied through GHG reduction investments outside the country or voluntary GHG emission reduction commitments of industries towards the national government.

- they are regarded as a long-term investment in business opportunities for the future. Short and medium-term losses are offset by expected long-term profits. For example, large private investors such as energy utilities will prefer their own projects since they expect these positive externalities. Small investors interested in reducing (unexpected) transaction costs, on the other hand, will prefer multilateral funds provided by a UN organization (Michaelowa & Dutschke, 1998, p. 17).

- they are made economically feasible through national programs.

In sum, there are various advantages to including JI and the CDM in national approaches to climate protection policy. Yet, there is still some reluctance on the part of Annex I Party private companies to participate in these flexible instruments. As a consequence, governments may provide additional incentives for private company participation; however, several questions about incentives for private company participation in JI and the CDM need to be taken into account.

Implications for the compensation of national energy/CO_2-taxes

In a number of Annex I countries, energy- or CO_2-taxes have been adopted (e.g., Denmark, Germany, Norway, and Sweden). Environmental policy in these countries enhance the incentives of private companies to participate in JI and the CDM through the compensation of CO_2-tax payments. Technically, there are no obstacles to offsetting a national CO_2-tax with emissions credits accruing from either JI or the CDM, since both instruments basically apply the same transaction unit (CO_2/CO_2-equivalent).

Though existing CO_2-tax systems do not include the other five GHGs, this is no obstacle for compensation. Proven and certified CO_2-emissions reductions from JI or CDM projects can directly compensate for domestic units of CO_2 or the five other GHGs. They also can be transferred into CO_2-equivalents and can offset domestic CO_2-emissions. The one major difference between JI and the CDM is that compensation from

Joint Implementation can occur at the earliest in 2008 whereas the CERs from the Clean Development Mechanism could compensate for national tax payments from the year 2000.

Taken together, the scope of incentives provided by the compensation of national obligations are dependent on their framework conditions. For a private company, it is favorable to participate in compensation projects if

- the marginal GHG reduction costs in the project are lower than the (marginal) tax rate. The reduction costs have to include the transaction and information costs;
- the marginal GHG reduction costs in the project are lower than the marginal GHG emission reduction costs in the company. If there are other options for the company to compensate for the tax burdens—for example, the utilization of energy efficiency potential in the firm (which is basically intended by the government imposing the CO_2-tax)—the company prefers this. Even if domestic reduction costs are slightly higher than the JI/CDM reduction costs, it is very likely the company will prefer this option because transaction costs of domestic actions are extremely low and final costs of JI/CDM projects are uncertain;
- the companies are not exempted or are not expected to be exempted from the CO2-tax or are not subject to lower tax rates.

In other words, to be a real incentive for private companies, the reduction cost differences, including the tax, must be substantial. For example, in 1998, the CO_2-tax rate on CO_2-emissions in Finland was \$13.39/t CO_2, in Denmark \$14.44/t CO_2, and in Sweden \$48.06/t CO_2, with parts of industry having lower tax rates (Arndt et al., 1998, p. 190). Though there are wide ranges of specific reduction costs—compared with the average greenhouse gas reduction costs in AIJ projects of between \$0.1 and \$19 per ton of CO_2-equivalent (see Table 7.2)—afforestation and reforestation projects as well as flue gas management seem to provide incentives for compensating a CO_2-tax. These figures, however, must be interpreted conservatively since some of the projects have not yet been implemented and their costs, as a consequence, have only been estimated.

The incentives for energy efficiency projects by Finnish or Danish companies are relatively low since reduction costs are near tax rates in Finland and Denmark; for the more expensive measures such as fuel-switching, the reduction costs are even higher than the tax rates. If there are lower tax rates or tax exemptions for certain industries—for

Table 7.2. Greenhouse gas reduction costs in 67 AIJ projects (1998, in U.S. \$/t CO_2-equivalent)

Project type	Reduction costs	Average costs[1]
Afforestation	0.4–0.6	0.1
Reforestation	0.04–64.2	0.1
Forest protection	0.5–3.2	2.2
Energy efficiency	0.9–582.4	10.4
Fuel switch	13.2–175.7	19.0
Management of flue gas	0.01–169.9	0.01
Renewables	3.1–244.5	8.7

Source: Institut für Industriebetriebslehre und Industrielle Produktion (IIP), 1998, Annex E: 334, own calculations. (1) weighted with the emission reductions.

example, power generation in Denmark does not have to pay the CO_2-tax; tax reductions for commercial energy consumption are granted based on the difference between the taxable returns including export and the inputs including imports—incentives of compensation in these industries are lower, even zero. Generally, incentives for participating in the projects are enhanced by imposing a higher tax rate as in Sweden. But it is questionable whether Finnish or Danish industry would have accepted such a high tax. On average, domestic CO_2-emission reductions that compensate for the payment of the CO_2-tax are more expensive than the CO_2-tax rates and higher than the emission reduction costs of possible JI and CDM projects. According to data from the Swedish power plant manufacturer, Alstrom, the replacement of an old coal-fired power plant by a new coal power plant is 13–25 US\$/t CO_2, by small hydro plants 40–50 US\$/t CO_2, wind power and biomass 50–100 US\$/t CO_2 and by solar energy above 120 US\$/t CO_2.

From the government's perspective, providing incentives through the possibility of a tax compensation may, however, bring about lower CO_2-tax revenues:

- In the no-compensation case, the companies will invest in GHG emission reduction measures up to the point where, assuming perfect markets, their marginal reduction costs will equal their marginal tax burden (i.e., the tax rate). If the marginal reduction costs are above the tax rate, it is economically rational for the company to pay the tax rather than to invest in GHG emission reduction measures. *Ceteris paribus,* the reduction measure reduces GHG emissions and the tax burden. On the company side, investments in emissions reductions have to be balanced with tax savings. On the government side, the losses of tax revenues have to be balanced with the welfare gains by the internalization of the external effects of excessive GHG emissions.

- In the compensation case, the companies can substitute JI or CDM measures abroad for domestic GHG emissions reduction measures. Under cost-minimizing conditions, the companies will do that as long as the marginal JI/CDM-reduction costs are below their domestic marginal reduction costs and the marginal tax burden. The compensation reduces the tax burden by the (higher) domestic tax rate and increases the emissions reduction investment costs only by the (lower) marginal reduction costs of the JI/CDM measure. Compared with the no-compensation case for the company, the emissions reductions are at lower cost. If there is sufficient global potential for JI/CDM projects with marginal reductions below the least-cost opportunities in the company, which is very likely, the tax burden of the company and the tax revenue of the government will be even zero. The company will only have to bear the reduction costs. The tax relevant marginal reduction cost of the company now is that of the JI/CDM project which is below the tax rate at any stage. Thus, for the company, it is at no point economically rational to pay the tax rather than to invest in GHG emissions reduction.

If the marginal tax burden increases (through a rise in the tax rate), more JI/CDM projects become profitable, taking into account a potential CO_2-tax compensation. Hence, the likelihood that the company will offset its tax payments with JI/CDM is even higher. If the government does not want to lose its tax revenues completely, it has to restrict the compensation to a certain percentage of the company's emissions.

In other words, ecologically, the compensation is a disincentive for the industry to save energy and reduce GHG emissions domestically. From a global perspective and in terms of the national Kyoto commitment, however, this has to be balanced with the emissions reductions abroad. From the public budget point of view in the Annex I Party governments, the question arises as to whether it is a higher burden for the public budget to renounce a certain part of the tax revenues or to run a national program promoting JI and the CDM to reach the same GHG emissions reduction effect with these instruments.

Incentives for the compensation of obligations in voluntary agreements

Similar to the compensation of a domestic tax, incentives for participation in JI and the CDM can be provided by the possibility of offsetting domestic obligations from voluntary industry agreements to reduce GHG emissions.

In Europe, several long-term voluntary agreements have been concluded (Oberheitmann, 1998, p. 24). As detailed in chapter 5, for example, German industry signed a voluntary agreement with the federal government in 1995 (subsequently modified in 1996 and 2000) to reduce CO2-emissions (see also Hillebrand, Buttermann, & Oberheitmann, 1997, p. 17). Companies could profit from the connection of their voluntary agreement and JI/CDM by the enlargement of the obligation to six greenhouse gases (rather than just CO_2), and possibly stretching of the commitment to the 2008 to 2012 budget period. Currently, CH_4, N_2O, PFCs, HFCs and SF_6 together provide for about 15 percent of the total GHG emissions in Germany.

With sixty-six million tons reduction potential by 2010, these gases count for about one-quarter of the total German reduction obligation. If the assumed development is realized and the specific reductions of the other five GHGs are less expensive than the specific CO_2-emission reductions, this has the potential for providing substantial flexibility (Hillebrand, 1999, p. 8). Stretching the industry commitment period brings economic and ecological advantages. On the one hand, a radical program to keep the 1995 obligation of the German federal government to reduce CO_2-emissions 25 percent by 2005 would incur severe sectoral as well as broader economic disruptions, since the marginal reduction costs in some cases would exceed 1000 DM/t CO_2 (Hillebrand & Wackerbauer, 1996, p. 107). Lengthening the commitment period also provides time to develop new GHG emissions reduction instruments. At the same time, extending commitment periods from five to ten years would not have large negative environmental impacts from the global climate perspective (Hillebrand, 1999, p. 9).

From a national perspective, the compensation of a voluntary agreement is favored for at least one reason. In contrast to a tax, offsetting obligations from a voluntary agreement has the advantage that industry cannot use other domestic options such as paying for emissions rights. There are a number of compatibility conditions for the integration of the German voluntary agreement and the flexible instruments arising from the general requirements of the Kyoto Protocol and the individual instruments that may inhibit their use. One problem derives from the fact that voluntary agreements are signed by industrial associations rather than single companies. A single company's investment in JI or CDM would benefit not only the single company but the whole sector, thus providing incentives for free-riding, especially by small- and medium-sized companies. In the opinion of the Federation of German Industry, large companies do not consider this a big problem since they receive the greatest benefits from these investments in absolute terms. Depending on the market structure of the sectors (if there are only a few big companies) and development of the investments in JI and CDM (growing marginal reduction costs over time), however, this free rider problem becomes more important in the future.

Potential obstacles to the effective operation of JI and the CDM

Specific projects can only be taken as JI or CDM measures if they provide for the additionality of their environmental benefits. The next section discusses obstacles relating to additionality that may occur due to the incentives provided by no-regrets measures on their own and incentives provided by national governments to participate in JI or CDM projects. Moreover, there are potential barriers to participation in JI or CDM projects due to provisions in the agreements under the auspices of the World Trade Organization (WTO). As discussed in the following section, it must be decided whether projects incompatible with the WTO can be regarded as additional.

Additionality and the provisions of the Kyoto Protocol

No-regrets measures

The incentives of no-regrets measures have a huge impact on the determination of the environmental additionality of the project concerned. There is an incentive for private companies to undertake climate change mitigation measures (e.g., enhancing the energy efficiency in production technology) if they are economically feasible and at the same time claim profits in terms of creating JI or CDM emission reduction credits. In other words, these investments are either business-as-usual, or windfall profits. The problem of defining additionality here lies in the high probability that the environmental benefits accruing from these projects would have occurred anyway, which would be incompatible with the provisions of the Kyoto Protocol. The issue is the timeframe. Are environmental benefits accruing from a project undertaken now additional if they would occur anyway in three or four years? To determine additionality, two questions must be answered: (1) Is it an environmental benefit to have the emissions reduced three or four years earlier? and (2) Will the emissions really be reduced in three or four years and, if so, by whom?

The first question is easy to answer since every ton of greenhouse gas reduced is an environmental benefit. The difficulty in determining additionality of no-regrets measures is connected with the timeframe of the development in absence of the project. Due to the asymmetric distribution of information, there are disincentives for the investing, and host Party to reveal to the UNFCCC bodies when emissions would be reduced if this information is available. This principal-agent problem of no-regrets measures is not solvable. The same applies when trying to determine if a project constitutes no-regrets measures. There are the same incentives for the company to obscure the break-even point of the project.

If no-regrets measures are politically determined to be additional because they provide a large potential for GHG emissions reductions, projects have to be undertaken on trust, regardless of the control measures implemented. Thus, the most important requirement for additionality of no-regrets measures is a maximum degree of transparency. Since the provisions of the Protocol do not exclude these measures, approval of a project as a JI or CDM measure must take this into account. A final decision on the acceptance of no-regrets measures is still open in the Kyoto process. For national governments, no-regrets measures are the least-cost alternative in providing incentives for private company participation, since no or little public finance is necessary. However, the difficulties regarding the concept of additionality, especially in connection with problems of correctly defining business-as-usual and environmental additionality, may lead to a rejection of these projects because of an incompatibility with the provisions of Article 6 and 12.

Projects in national programs

In the recent years, the Netherlands had a budget for AIJ of $12 million. In 1998, the government provided an additional $50 million for JI projects and capacity building measures. From 1999 to 2003, a new budget of $250 million for the CDM and $150 million for JI has been allocated. Denmark's JI program will be launched soon, but on a much smaller scale (three to five pilot projects in the Baltic states). The Norwegian business sector is very interested in JI projects in Eastern Europe with a domestic permit trading system likely to start within the next years. Sweden established the Swedish Program for an Environmentally Adapted Energy System with a volume of about 330 million SEK ($42 million) up to 1998 to finance projects on energy efficiency and increased use of renewables in the Baltic Region and Eastern Europe (Energimyndigheten, 1998, p. 1). Currently, about seventy Swedish AIJ projects in the Baltic region are operating. These projects cover the whole range of project types from energy efficiency measures to afforestation.

The incentives for projects to be included in national programs may have an impact on the concept of additionality under certain conditions. Projects may be economically feasible for private companies if they are run through a national program or are directly subsidized, thereby reducing the investments of the company or possibly promoting additional exports of technology into the host countries. These are powerful incentives for private companies to participate in JI and CDM projects. The environmental benefits accruing from projects financed through a national program are, in essence, additional if they have been launched with public finance, since they are not economically feasible for private companies. The difficulty in determining additionality here basically lies in the feasibility of the single projects. As mentioned above, however, there are certain difficulties in determining the feasibility in no-regrets measures. If national climate policy chooses projects other than no-regrets measures to support, there are no conflicts with the concept of additionality deriving from this incentive. Nevertheless, projects in national programs may face difficulties with the provisions of agreements governed by the WTO regarding export subsidies.

Conclusions

As detailed in this discussion, there are various advantages to including JI and the CDM in national approaches to climate protection policy, paramount among them being the potential for lowering the cost of meeting national GHG reduction targets. Yet, there is currently reluctance on the part of Annex I Party private companies to participate in the flexible instruments of the Kyoto Protocol. As a consequence, national governments may find it necessary to provide incentives that encourage private sector participation in these mechanisms: for example, national programs that would reduce the financial risk for companies investing in projects abroad or compensation for domestic CO_2-taxes or voluntarily agreed reductions in GHG emissions by industry with emission credits from JI or the CDM. Providing these types of incentives, however, has implications for national policy.

One such implication, the provision of incentives by the compensation of a national CO_2-tax, may lead to lower tax revenues for the government. If there are no limits imposed on compensation, CO_2-tax revenues may even approach zero. As a consequence, this type of compensation serves as a disincentive to reduce GHG domestically. Seen from a global environmental perspective, however, the location of the

CO_2 emissions reductions is unimportant. Moreover, the tax losses may be offset by the welfare gains from lower external costs.

For another, where projects may be economically feasible (no-regrets) but are not undertaken due to political or economic framework conditions in the host country, national governments may provide incentives for participation in such projects through national programs. Such incentives, however, have a huge impact on the determination of the environmental additionality of the project concerned. The problem of defining additionality here lies in the high probability that the environmental benefits accruing from these projects would have occurred anyway and thus would not be accepted by the FCCC bodies. By promoting such measures, especially those financed in national programs, governments risk spending public money on projects that may have been undertaken by industry anyway.

References

Albrecht, J. (Ed.). (2002). *Instruments for climate policy: Limited versus unlimited flexibility.* Ghent, Belgium: University of Ghent.

Arndt, H-W., Heins, B., Meyer, E. C., Pfaffenberger, W., & Ströbele, W. (1998). Ökosteurn auf dem Prüfstand der Nachhaltigkeit. Berlin.

Capros, P., Mantzos, L. (2000). *The economic effects of EU-wide industry level emissions trading to reduce greenhouse gases.* Athens: Author.

Energimyndigheten. (1998). Climate Related International Energy Cooperation—The Swedish Programme for an Environmentally Adapted Energy System (EAES): Projecs on energy efficiency and increased use of renewable energy sources in the Baltic Region and Eastern Europe. Progress Report 1998. Eskilsatuna: Swedish National Energy Administration.

Framework Convention on Climate Change (FCCC). (1995, June 6). *Report of the conference of the Parties on its First Session, Held in Berlin from 28 March to April 7 1995. Addendum Part Two: Action taken by the Conference of the Parties at its First Session.* FCCC/CP/1995/7/Add.1.

Framework Convention on Climate Change (FCCC). (1998, March 18). *Report of the Conference of the Parties on its Third Session, Held at Kyoto from 1 to 11 December 1997.* FCCC/CP/1997/7/Add.1.

Framework Convention on Climate Change (FCCC). (2001). *Report of the Conference of the Parties on its seventh session, held at Marrakesh from 29 October–10 November 2001. Part 2: Actions taken by the Conference of the Parties.* FCCC/CP/2001/13/Add.1 and Add. 2. Volumes 1 and 2.

Hillebrand, B. (1999). *Auswirkungen nationaler und internationaler Klimapolitik auf die Zementindustrie. Zement, Kalk, Gips International.* Wiesbaden.

Hillebrand, B., Buttermann, H.-G., & Oberheitmann, A. (1997). *CO«-Monitoring der deutschen Industrie—ökologische und ökonomische Verifikation.* Untersuchungen des Rheinisch-Westfälischen Instituts für Wirtschaftsforschung,

Heft 23/1, Essen. Rheinisch-Westfälisches Institut für Wirtschaftsforschung.

Hillebrand, B., & Wackerbauer, J. (1996). Ökologische und ökonomische Wirkungen von CO_2-Minderungsstrategien. *RWI-Mitteilungen, 47*(1–2), 107–131.

Institut für Industriebetriebslehre und industrielle Produktion (IIP). (1998). *Zur Effizienz einer länderübergreifenden Zusammenarbeit bei der Klimavorsorge—Analyse von Joint Implementation unter Einbeziehung eines Emissionsrechtehandels für die Bundesrepublik Deutschland, die Russische Föderation und Indonesien.* Untersuchung des Instituts für Industriebetriebslehre und industrielle Produktion (IIP), Universität Karlsruhe.

Michaelowa, A., & Dutschke, M. (1998). Creation and sharing of credits through the Clean Development Mechanism under the Kyoto Protocol. Paper presented on the experts workshop "Dealing with carbon credits after Kyoto," 28–29 May 1998, Callantsoog, The Netherlands.

Michaelowa, A. (1998a). AIJ cannot function without incentives. In P. Riemer, A. Smith, & K. Thambimuthu (Eds.), *Greenhouse gas mitigation—Technologies for activities implemented jointly.* Amsterdam, 403–408.

Michaelowa, A. (1998b). Climate policy and interest groups—a public choice analysis. *Intereconomics, 33*(6), 251–259.

Oberheitmann, A. (1998). *European Union energy policy and recent energy policy proposals relevant to climate change.* RWI-Papiere, Nr. 53, Essen. Rheinisch-Westfälisches Institut für Wirtschaftsforschung.

Oberheitmann, A. (1999a). On the incentives for undertaking JI- and CDM-measures through the compensation of national CO_2-tax payments. In Bundesministerium für Umwelt, Naturschutz und Reaktorsicherheit (Eds.), *Reports on AIJ projects and contribution to the discussion of the Kyoto mechanisms* (pp. 100–104). Bonn: Bundesministerium für Umwelt, Naturschutz und Reaktorsicherheit.

Oberheitmann, A. (1999b). Some remarks on additionality and incentives in activities implemented jointly, Joint Implementation And The Clean Development Mechanism. RWI-Papiere, Nr. 59, Essen. Rheinisch-Westfälisches Institut für Wirtschaftsforschung.

CHAPTER EIGHT

The Costs and Benefits of Emissions Trading

GARY C. BRYNER

Introduction

Environmental policy in the United States is a study in paradox. Public opinion polls and other measures of public sentiment show strong support for environmental regulation (Dunlap, 1992). Recent reports have found that significant progress has been made towards improving environmental quality; the nation's air and water is cleaner; national programs help implement new international agreements to reduce global environmental risks; threats to children's health posed by lead in the environment have diminished; farmers are reducing soil erosion and pesticide runoff into streams; and industries are reducing their release of toxic pollutants (U.S. Council on Environmental Quality, 1995).

Despite this progress, reforming environmental law and regulation has become a regular refrain of American politics and policymaking. As soon as the ink was dry on the first modern federal environmental laws, critics began questioning whether they were too expensive and intrusive. As the framework of environmental law developed into a complex maze of statutes, rules, and agencies, a growing number of voices, from a variety of perspectives, have called for changes in the ways we regulate pollution and pursue environmental quality.

One of the most important and popular remedies for the ills that many believe plague environmental regulation is the use of market-based approaches to regulation. In the conventional—or command-and-control—approach to regulation, federal agencies issue national standards that apply to all sources in all areas, and agency officials mandate specific technologies and compliance actions. Market-based regulatory tools seek to create incentives for companies to find the most effective ways of reducing emissions. This gives companies the flexibility to devise their own methods for achieving the required reductions.

As discussed elsewhere in this volume, there are several kinds of market-like mechanisms that might be employed in environmental regulation (U.S. Congress, Office of Technology Assessment, 1995). The two most controversial instruments are pollution charges and emissions trading. Pollution charges, or taxes, are levied on emissions or on input to activities producing pollutants. Polluters are charged a fee for each unit of pollution they emit. Pollution taxes or fees, if high enough, provide strong incentives for companies to reduce emissions in whatever way is most efficient for them–closing down some operations, using cleaner fuels, investing in control technologies, changing work practices, and so on. Pollution charges provide clear incentives for reducing emissions below permitted levels.

In emissions trading, polluters are allocated a limited number of allowances or units of emissions for release into the environment. Companies can either make the changes necessary to stay within the limits, or they can buy allowances from others. Tradable emission permits can be saved (or banked) for future use, or sold by polluting companies as long as limits on total emissions are not exceeded. Polluters have incentives to reduce emissions beyond the allowances given to them so that they can generate revenues through the sale of excess allowances (Hahn, 1988). Over time, total emissions can be reduced further by decreasing the number of allowances distributed to pollution sources.

The use of market-based regulatory tools raises a number of important questions about environmental policy. I focus on three here:

1. How well do market-based approaches help achieve their goal of reducing the costs of compliance and making regulation more cost effective?

   Given the billions of U.S. dollars spent on pollution control and cleanup ($140 billion a year according to some estimates) there are strong expectations that market-based tools can significantly reduce the costs of environmental regulation.

2. To what extent do these regulatory tools contribute to the achievement of environmental quality goals?

   No matter how well market instruments reduce the costs of pollution control, the primary question is whether environmental problems have been effectively remedied and natural resources protected.

3. What are the implications of the experience of emissions trading in the United States for a greenhouse gas or carbon trading program?

   A carbon trading program has been a prerequisite for U.S. support of an international climate change agreement, and many businesses have entered into voluntary early reduction programs in anticipation of an eventual GHG trading program.

Emissions trading programs are the most popular market-based regulatory tools the United States government is experimenting with because of the potential to reduce compliance costs, increase the flexibility given regulated industries, improve the efficiency and effectiveness of regulatory agencies, and achieve environmental and public health goals. It is for these reasons they are the primary focus of this chapter. Emissions trading programs have been enthusiastically embraced by both regulated industries and policymakers and have become part of virtually every policy proposal currently before Congress and state and federal environmental protection agencies. It is difficult to imagine a new regulatory program that does not include at least some form of emissions trading. This chapter reviews the experience with three kinds of trading programs and proposals:

1. federal emissions trading programs under the federal Clean Air Act, including sulfur dioxide emissions in the acid rain program, and other national efforts, which were the earliest efforts to devise emissions trading programs;

2. state innovations, such as Southern California's South Coast Air Quality Management District's RECLAIM program for trading sulfur and nitrogen oxide emissions;

3. proposals for trading of GHGs.

This mix of regulatory innovations at different levels of government, for different environmental problems, and for different countries provide a broad base for examining the advantages and disadvantages of market-based regulatory tools.

Emissions trading programs are quite new, so definitive studies of the costs and benefits will come in the future. Given the tremendous interest in them, this study assesses the strengths and weaknesses of trading programs, based on the limited experience with them, and suggests questions that policymakers and others should ask before they embrace market-based tools, particularly emissions trading, as a way to improve the effectiveness of existing regulatory programs and design the next generation of environmental laws and policies. Emissions trading programs are also used to deal with other environmental problems such as reducing water pollution and the generation and storage of hazardous wastes, but space does not permit a discussion of all trading programs here (Wirth & Heinz, 1988, 1991; Dudek & Palmisano, 1988; Stavins, 1989). The experience of emissions trading in air pollution programs provides important lessons for its use in addressing other pollution problems.

An overview of traditional emissions trading programs in place

Federal government emissions trading programs

The United States Environmental Protection Agency (EPA) has experimented with a number of emissions trading programs. In 1986, the EPA issued an Emissions Trading Policy that outlined guidelines for several emissions trading programs: bubbles, netting, emission reduction credits (ERCs), and banking of ERCs (51 FR 43814, December 6, 1986; 40 CFR 51, Appendix X). EPA's Emissions Trading Policy allows sources to create ERCs only if the reductions are surplus, enforceable, permanent, and quantifiable. Surplus emissions are not required by government regulation; enforceable emissions are approved by a state and subject to EPA enforcement actions; permanent reductions are those that legally prohibit the resumption of emissions; and emissions must be quantified before and after the reduction. Trades can only be for the same pollutant, must satisfy applicable monitoring requirements, and must be approved by the EPA as part of state implementation plans. The different emissions trading options under federal clean air laws and regulations include the following:

Offsets	New sources beginning operation in areas that have not met national air quality standards buy enough emission reductions to offset the new source's projected emissions, plus an additional percentage of reductions so that the new source helps contribute to the attainment of the air quality standards.

Netting	If new emissions come from expansion of an existing source, the increase can be offset from internal sources, through netting, so that the facility's owners can avoid the costs and delays associated with obtaining permits to operate new sources of pollution.
Bubbles	Companies can receive credit for emissions reductions in some areas for higher emissions elsewhere; total emissions from each facility are viewed as encapsulated within a large bubble, rather than from individual smokestacks.
Averaging	Producers of products being phased out, such as leaded gasoline and heavy duty engines, have been allowed to show compliance by using an average of their products, rather than being required to show that every product complies with the standard.
Early reduction	Sources of pollutants can gain credits or delay compliance with future standards if they reduce emissions before reductions are required.
Cap and trade	Allowances are allocated to sources; a cap on total allowances is set and ratcheted down to meet environmental goals; sources can buy, sell, or trade allowances in order to meet their emission limits.

Beginning in 1974, the EPA instituted its first trading program, the Offset Program. The Offset Program required new sources beginning operation in areas that have not met national air quality standards must buy enough emission reductions to offset their projected emissions, plus purchase an additional percentage of reductions in order to contribute to the attainment of the air quality standards. Most of the credits purchased for offsets have come from closing down existing facilities. One study estimated that more than 10,000 tons of pollutants have been bought and sold through offsets for more than $2 billion, but it is difficult to obtain information because these trades are private transactions (Korb, 1998, p. 107).

If new emissions come from the expansion of an existing source, the increase can be offset from internal sources through a process called netting. Netting has become the most commonly used trading program because it has fewer restrictions than offsets. Owners of existing facilities frequently construct new sources of pollution, and they use the netting process to avoid the costs and delays associated with obtaining permits to operate new sources of pollution. One of the EPA's most recent innovations, the Environmental Excellence and Leadership Program or XL program, includes pilot projects that allow facilities to alter their mix of emissions without having to obtain a new operating permit as long as total emissions do not exceed the facility's cap (Korb, 1998, p. 105). The second EPA trading program, proposed in 1979 and put in place in 1986, was the bubble policy. This policy allows companies to receive credit for emissions reductions in a specified area, as opposed to an individual location. Total emissions from each facility are viewed as encapsulated within a large bubble, rather than from individual smokestacks, or two, new adjoining power plants use the bubble concept to combine emissions. Regulatory officials established maximum total allowable emissions and left managers free to determine optimal emissions from individual sources, so that total compliance costs could be minimized. By the mid-1980s, the EPA approved some fifty bubbles and states had approved many more (under authority given them). Bubbles are still used to give facilities flexibility in meeting emission limits (Korb, 1998, pp. 108–109).

A third experiment (averaging) with trading began with the phaseout of lead in gasoline in 1979 when the EPA limited the average lead content permitted for large refiners. Refiners could average the content of leaded and unleaded gasoline in order to show compliance. In order to meet goals for further reductions required by 1982 and beyond, the EPA allowed refiners and importers of fuels that reduced the lead content of their fuel below the new EPA standards to sell the credits earned to other refiners or importers by allowing them to average the lead content of leaded and unleaded fuel. Banking was discontinued in 1987, when all fuels were required to meet the standard, and the program was terminated in 1996 when all leaded gasoline for highway use was banned.

Economic Incentive programs (EIPs), such as emissions fees or marketable permits, can be used by states in certain areas that have not attained national air quality standards, and are required in states that fail to show they are making reasonable further progress in addressing serious carbon monoxide or extreme ozone problems (42 U.S.C. 7511a(g)3). The EPA issued rules for EIPs in 1994, which provide that credits for emission reductions for trades, sales, or offsets can count as EIPs (59 FR 16690, April 7, 1995, 40 CFR part 51). In 1995, the agency proposed a model open market trading rule to guide states as they create emission reduction credit programs (60 FR 39668, August 3, 1995). These are primarily used in states by new sources of pollution to satisfy the offset requirement by purchasing credits from existing ones.

The Stratospheric Ozone Protection Program established a marketable permit system for producers and importers of chlorofluorocarbons

(CFCs). Under this program, the EPA allowed manufacturers of heavy duty engines to meet emission standards by averaging emissions across their entire production. The program also allows averaging between manufacturers. The Air Toxics Early Reduction Program allows sources of toxic pollutants that reduce emissions by 90 percent before national emission standards for hazardous air pollutants are proposed, to be able to delay their compliance with those standards when they are eventually issued (Korb, 1998, p. 105).

The EPA has also mandated emission limits for nitrogen oxide emissions from electric utilities in the twelve northeastern states and the District of Columbia in an effort to reduce ozone pollution that has been particularly difficult to reduce. The Ozone Transport Commission, established by the 1990 Clean Air Act, authorized states in the Northeast to address NO_x emissions as a regional air problem. A 1994 agreement between states established a trading program for emission allowance trades for sources located in different states (63 FR 25902, May 11, 1998). The model rules for trading NO_x allowances in the Ozone Transport Region were issued by the EPA in 1998. States are not required to engage in trading, but if they do adopt the rule as part of their implementation plan, the EPA promises to approve it quickly (63 FR 57356, October 27, 1998). The emissions trading program is expected to spread to twenty-two states in 2003. An interstate emissions trading for the OTC states, and an additional twenty-four eastern state, is also being developed (U.S. General Accounting Office, 1998, p. 4). Allowances are issued for use in a particular year but can be banked for future use. The EPA maintains an allowance tracking system that gives a serial number to each allowance in order to track it through trading. The trading scheme is quite complex and has been the target of legal challenges (Holtcamp, 2000).

One of the most important innovations of the Clean Air Act of 1990 was the market-based incentive system to reduce acid rain-producing emissions from coal-fired power plants through a nationwide cap and trade system. The first set of major rules to implement the sulfur dioxide (SO_2) allowance trading program were issued in 1993, and took effect in 1995. The heart of the acid rain emissions trading system is a cap on total emissions projected, by the year 2010, to result in a reduction of SO_2 emissions of ten million tons from 1980 levels. The plan is implemented in two phases. In Phase I, one hundred and ten plants in twenty-one Midwestern and eastern states were given allowances (a permit to emit one ton of SO_2). Allowances have been allocated for the years 1995 through 2030. In Phase II, beginning in 2000, the number of sources included in the program will be greatly expanded to include all large fossil fuel electric utility generating units. The allowances for these sources are allocated by a formula based on the amount of fuel the plants used during a base year period. The number of allowances will be capped at eight point nine five million tons per year by 2010 when the program is fully implemented. Plants able to control emissions below the levels allocated to them can bank them for future use, trade them, or sell their excess emission permits to others who exceeded their allowance. Transactions need not be approved by the EPA; the agency auctions a small fraction of the total allowances allocated each year at an annual auction, in order to provide for growth in new sources. The first auction of SO_2 allowances took place in 1993. Anyone who wishes can bid for allowances, and all allowances are sold each year. Participating sources must meet stringent monitoring requirements, usually through continuous emissions monitoring.

State and local government emissions trading programs

There is considerable variety across the states in terms of the specific elements of emissions trading programs (U.S. Environmental Protection Agency, 1999). Some fourteen states have developed programs designed to reduce air pollution that include variations on the idea of emissions trading. These programs were largely created between 1994 and 1998. The purpose of these state-level programs is primarily to help bring the areas into attainment of national ambient air quality standards. These pollutants are particulate matter (PM), ozone, sulfur dioxide (SO_2), nitrogen oxide (NO_x), lead, and carbon monoxide. Emissions of lead, the other pollutant regulated through national air quality standards, have fallen so dramatically as a result of unleaded fuel that lead is no longer a major concern of air pollution regulation.

Most of these state programs are voluntary; regulated sources need not bank, sell, or otherwise trade in emission allowances. If sources choose to participate, they typically must demonstrate reductions used to claim credits are quantifiable, real reductions in emissions, subject to enforceable requirements, and surplus, and represent permanent reductions. Credits usually have a ten-year life time. Some programs are limited to stationary sources, such as power plants and factories, while others include the creation of credits through vehicle scrappage programs. Some programs include reductions from area sources such as off-road equipment, consumer products, small commercial facilities, and other dispersed sources. Some states allow sources themselves to verify the creation of surplus allowances, while others require independent verification. Different states allow or prohibit inter-pollutant trading. Many states build into their trading program a requirement that the value of credits be reduced or discounted in order to decrease emissions over time and ensure that air quality improves (U.S. Environmental Protection Agency, 1999).

One of the most ambitious emissions trading schemes has been developed in southern California. The South Coast Air Quality Management District (SCAQMD), the agency responsible for addressing the greater Los Angeles area's air pollution problems, put in place an emissions trading program called the Regional Clean Air Incentives Market or RECLAIM in 1994. RECLAIM affects some 350 sources in the Los Angeles area that produce at least four tons of nitrogen or sulfur oxide emissions a year. This represents about 65 percent of the NOx and 85 percent of the SOx emissions from stationary sources in the region. These sources are required to reduce their emissions by a fixed percentage each year for the regulated pollutants. They are given an annual allocation of allowances and are free to find the most cost-effective means to reduce emissions. Sources that reduce NO_x and SO_2 emissions beyond the caps imposed on them may sell their excess credits, called Regional Trading Credits, to companies that have exceeded their limits (U.S. General Accounting Office, 1997).

Other countries' experience with emissions trading programs

As is true in the United States, European countries primarily rely on traditional or command and control approaches to environmental regulation. Most of the advanced, industrialized democracies of Western Europe—countries that are members of the Organization of Economic Co-Operation and Development (Canada, Japan, and the United States are also members of OECD)—have increasingly devised and implemented market-based innovations to regulation, but have selected approaches much different than the emissions trading programs employed in the United States. The kinds of market-based instruments in European environmental regulation include water effluent charges, eco-taxes, deposit-refund systems, and emissions trading.

Few European countries have followed the United States' lead in developing tradable emissions permit systems. German air pollution law allows sources to transfer emission reduction obligations but that only rarely occurs. The Netherlands has a bubble policy for power plants which allows facilities to treat all emissions from one facility as one source. Australia, Canada, and the United Kingdom have considered pollution trading policies, but no country has embraced emissions trading except the United States. The primary focus in Europe has been on green taxes, rather than on trading programs. The United

States is one of the few countries where tradable permits have been used extensively. Its experience deserves particular study.

Results of emissions trading programs

Emissions trading is an important, useful innovation in environmental regulation in the United States. As the experience with this regulatory tool in federal and state environmental programs shows, pollution trading reduces compliance costs of regulated industries and gives them more flexibility to meet emission goals. Emissions trading also helps generate political support for new regulatory programs and serves as a basis for fashioning compromises acceptable to a wide-range of interests. Carefully designed programs can simplify some of the regulatory tasks of government agencies and the compliance burdens on regulated sources. Trading programs may reduce the cost of achieving expensive environmental goals to acceptable levels, and may make it possible to take on environmental problems that otherwise are not addressed.

Reducing regulatory costs

Emissions trading programs generally reduce the cost of compliance for regulated industries. The acid rain control program is widely viewed as resulting in a significant reduction in the cost of pollution controls. The initial projections for meeting the goals of the acid rain program estimated compliance costs of $4.9 billion a year by 2010. The U.S. General Accounting Office's (1997, p. 3) most recent estimate of those costs is now less than $2 billion a year in 2010. A study by Resources for the Future concludes that the expected benefits of the program far outweigh the costs of compliance because of the flexibility it gives to utilities to find the cheapest ways to reduce emissions and to take advantage of trends in fuel prices and transportation costs (Burtraw, Krupnick, Mansur, Austiin, & Farrell, 1997).

Industry groups estimated that the cost of reducing a ton of emissions under the traditional regulatory approach at about $1,500/ton; the EPA's estimate was about $650/ton. The actual prices of allowances available for purchase between 1993 and 1996 at the Chicago Board of Trade, where allowances are traded like other commodities, fell from $122 to $66. Transactions for more than thirty-six million allowances have been recorded, twenty-five million of which have been traded between different companies. Trades within companies have been much greater than those between firms but the actual number of trades, while substantial, has been lower than some analysts had projected. Companies have been less interested in using trading to reduce compliance costs than they have been in banking allowances for use in Phase II of the program, when requirements are more stringent (Bohi & Burtraw, 1997, pp. 5–6). The EPA's decision to eliminate its case by case review and approval of each trade has also reduced the administrative and transaction costs. The program costs about $12 million a year to operate, which translates into an administrative cost of $1.50/ton of pollution reduced. The EPA has also auctioned over 500,000 allowances since 1993 (Korb, 1998, p. 112).

According to a 1998 study of the U.S. National Science and Technology Council (1998, p. 93), the costs of administering the acid rain trading program are lower than traditional regulatory schemes because it "eliminates the need to devise source-specific emission limits and to review control technologies and detailed compliance schedules." Lower costs are attributed to "cost reduction efforts and improved performance of scrubbers and changes in fuel markets;" it is difficult to estimate "future technological improvements, the more efficient use of existing technologies, and future economic conditions," but innovation, prompted by competition, seems to have fueled cost savings (U.S. National Science and Technology Council, 1998, p. 93).

Other studies confirm that much of the cost savings was a result of switching to low sulfur coal. Six states in which the power plants regulated under Phase I of the program are located have regulatory programs that favor scrubbing as a compliance option, as a way to protect their local coal industries. Fourteen facilities operating in these six states opted for scrubbing as a result of regulatory pressures. Installation of scrubbers was also encouraged by awarding bonus allowances for utilities that met their emission reduction targets through scrubbing. The cost of scrubbing also fell in response to competition from other means of reducing emissions, such as fuel switching. Overall, low sulfur fuel was responsible for about 50 percent of the emissions reductions achieved in Phase I. The costs of cleaner coal were lower than projected because of the deregulation of the railroad industry, the development of fuel-mixing technologies that allowed high and low sulfur coal to be mixed together to reduce emissions, and the surprising decline in low sulfur coal prices even as demand increased. The cost of switching from high to low sulfur coal has been about 35 percent of the cost of installing scrubbers. Scrubbing may become more necessary to comply with the more stringent Phase II requirements, and one of the major effects of the trading program may be to delay the installation of scrubber and other controls (Bohi & Burtraw, 1997, pp. 13–19).

Southern California's emissions trading program has been successful in achieving its goal of reducing the cost of reducing air pollution. A 1996 audit by the South Coast Air Quality Management District found that during its first two years, more than 100,000 tons of nitrogen and sulfur oxides had been traded for more than $10 million, and that the trading program was on track to meet its goal of reducing NO_x emissions by seventy-seven tons/day and SO_x emissions by fifteen tons/day by 2003. The price of NO_x credits in 1994 were $24/ton and $132/ton for SO_x, well under the $15,000/ton level that is the state of California's trigger point for reviewing the program's cost effectiveness (South Coast Air Quality Management District, 1996). During the following fifteen months (1996 and the first quarter of 1997), more than $20 million worth of credits were traded. By 2000, the prices of NOx credits in the RECLAIM program had increased twenty times from 1998 levels, and reached as high as $8,000/ton, indicating a shrinking supply of credits (Environment NOW News Service, 2000). In the Northeast, the NO_x allowance market is quit new. In 1999, prices of allowances varied from $710 to $7,600 per allowance (Airtrends, 1999).

Emissions trading may not reduce compliance costs in every case since not all emissions trading programs are relatively simple. The EPA's bubble program is a complicated system that requires the agency to review each transaction. Transaction costs are relatively high, often $10,000, according to one estimate, which may be counterproductive for small trades (Korb, 1998, p. 107). It requires extensive emissions modeling, assessments of alternative control technologies, compliance schedules, and is part of a complicated system of command-and-control emission limits and technological controls. What seems to work best is allowing trading within a facility, so that plant managers do not need to obtain new operating permits when production processes change, as long as the total emissions do not exceed its cap.

Achieving environmental goals

Emissions trading programs can be effective tools for improving environmental quality and preserving natural resources if sufficient information is available to policy designers to ensure the emission targets they set will accomplish the environmental goals. If the targets are sound, and the trading program is effectively monitored and enforced, then the goals will likely be achieved. However, the national and state trading programs in place have not yet resolved the problems at which they were aimed. While it is too early to judge emissions trading definitively, there are some troubling indicators.

Under the acid rain program, sulfur dioxide emissions from the power plants in Phase I of the program (1995–1999) were to decline each year for a total reduction of three and a half million tons; during Phase II (2000–2010), emissions are to fall another five million tons. Emissions in 1996 were 35 percent below the cap for that year. The program has reduced emissions and may prevent future damage, but emission reductions do not appear to be sufficient to restore lakes in the Adirondacks and elsewhere that are highly acidified. Lakes vary in their susceptibility to acid deposition, and some lakes are more damaged than others. The program's goal of reducing total emissions does not appear to be sufficient to provide protection for all lakes. Legislation before Congress in 1999 called for a study to assess whether emissions reductions projected under the current acid rain program are sufficient to protect lakes in the Adirondacks and elsewhere (S. 172, 106th Congress, 1999). A U.S. General Accounting Office report issued in March 2000 found that while sulfur levels had declined significantly in the Adirondacks, nitrogen levels have continued to rise, and the resultant nitric acid threatens fish larvae and adult fish (U.S. General Accounting Office, 2000). In response to this and other studies warning that the acid rain program may not be sufficient to allow the recovery of ecosystems in the Northeast, the New York State legislature passed a law in May, 2000, prohibiting New York firms from keeping any money they receive by selling emission credits they earn from reducing emissions to upwind sources in the Midwest and South (Hernandez, 2000).

In August 1997, the California Air Resources Board, a state agency, suspended all rules that permitted the trading of VOC emissions because of complaints that the trading program had violated federal civil rights laws. The complaint argued that emissions trading in the Los Angeles area between stationary sources and motor vehicles resulted in disproportionately higher air pollution levels in minority communities near the facilities that purchased the credits (Stevens, 1997). In theory, such problems are not possible, since the Clean Air Act requires monitors to be located throughout airsheds and compels state officials to show attainment with the national standards at all monitors. But research suggests national standards for pollutants like particulate matter and ozone are not sufficiently strong to protect humans from adverse health effects, so higher concentrations in minority communities, even if national standards are not exceeded, can be unfair. Other problems of inadequate monitoring, the wide distribution of mobile source emissions and more concentrated ones from stationary sources, the intermingling of toxic air pollutants and the conventional ones regulated by ambient air quality standards, all pose environmental justice-related problems for emissions trading (Natural Resources Defense Council, 1996).

Emissions trading programs are a cost-effective means to accomplish the goals of environmental policies, but the goals themselves may not be sufficient to remedy the problems at which the policies are aimed. This is not the fault of emissions trading programs; they are in a sense policy tools that are neutral in terms of policy goals. If regulatory programs are not carefully designed to ensure that all areas meet minimum air quality standards, trading systems may not produce the environmental and health benefits expected of them.

Broader issues in assessing emissions trading

Trading programs clash with some important expectations held for environmental policy. Trading schemes are, in one sense, inconsistent with the polluter pays principle that is one of the key values underlying environmental regulation. Trading tends to distribute equally the cost of pollution controls across all sources, rather than imposing the greatest control costs on the sources that produce the greatest emissions. Some firms are still able to externalize some of their costs of production to other sources, rather than ensuring they account for all of those costs (Seligman, 1994). Trading programs may not reward sources that have reduced emissions voluntarily. If allowances are distributed based on past emission levels, then high polluters will, in effect, be rewarded for their recalcitrance, while innovators who have already invested in emissions reductions will have fewer allowances to work with.

One of the underpinnings of environmental policy has been to encourage or force the development of cleaner, less polluting technologies. Despite the flaws in the conventional approach to regulation, it has often served to expand the use of cleaner technologies and to encourage the development of new technologies. This momentum can be lost if firms find emission credits available at a lower price rather than investing in newer, cleaner technologies. However, trading gives sources more flexibility in meeting standards, and they are able to make more use of cheaper technologies. For example, trading encourages washing coal to reduce its sulfur content, a technology that alone will not lead to the achievement of emissions standards, but can be combined with other actions, such as purchasing credits, at a fraction of the cost of other control technologies (meeting with the author, August 14, 1999).

In contrast to emissions trading, pollution taxes provide a continuous incentive to devise new processes and technologies, since every time reductions are made, lower taxes result. Emissions trading programs, unless aggressively structured, may reduce the pressure for developments in technology. Trading programs, unlike pollution prevention efforts, simply move pollution from one location to another rather than provide clear incentives to reduce emission levels. Credits earned by plant shutdowns create incentives for regulated industries to close existing facilities and move to new, less regulated areas. Accurate past emissions from these sources may be difficult to obtain.

As the Los Angeles case shows, emissions trading programs may not adequately remedy the public health threats posed by pollution. They assume there are safe levels of pollution for human health and ecosystems, but estimates of safe thresholds of exposure are often difficult to make. They result in increased exposures to some groups, particularly minority communities, who live near areas where sources find it cheaper to buy credits from others rather than reduce their emissions. Residents who are poor or lack the political clout to demand pollution reductions may be exposed to higher levels of toxic pollutants than their wealthier neighbors, and higher levels than before trading programs were instituted (Drury, Belliveau, Kuhn, & Basnal, 1999, pp. 251–258). They encourage the diffusion of relatively inexpensive control technologies that reduce emissions enough to achieve environmental standards when combined with the purchase of emission credits. Most importantly, voluntary emissions trading serve as a transition to and help build support for a strong regulatory program that employs carbon trading and other market-based regulatory instruments in reducing GHG emissions to a safe level.

Emissions trading: some lessons for climate change

The United States and some 175 other nations have signed the United Nations Framework Convention on Climate Change (FCCC) that was announced at the 1992 Rio Earth Summit. The U.S. Senate subsequently ratified the convention, committing the nation to the nonbinding target of limiting greenhouse gas emissions (GHG) at the year 2000 to 1990 levels. The United States adopted the Berlin Mandate in March 1995, at the First Conference of the Parties (COP-1) in Berlin, Germany, an agreement that structured future negotiations and provided that developing countries would initially not be required to make binding greenhouse gas reduction commitments (Flavin, 1995; *Nature*, 1995). In July, 1996, at the Geneva Climate Summit, the United States committed to legally binding targets and timetables for reducing GHG emissions in the more developed world (Environmental Defense Fund, 1996).

The Kyoto Protocol signed in December 1997 required thirty-eight industrial nations to reduce their emissions of six greenhouse gases by about 5 percent from 1990 levels between 2008 and 2012. The United States agreed to reduce greenhouse gas emissions by 7 percent; Japan agreed to a 6 percent cut, and the European Union, 8 percent (Oberthuer & Ott, 1999). The Clinton administration announced it would not submit the treaty to the Senate for ratification until there was both meaningful participation from developing nations, and a clear agreement that flexible measures (primarily market-based mechanisms such as GHG trading, joint implementation, and CDM that allowed industrial countries to offset their emission reductions with investments in energy efficiency and other projects in emerging countries) could be used to demonstrate compliance. In November 1998 during the Buenos Aires talks on climate change, the Clinton administration signed the Kyoto Protocol, in an effort to give a boost to the flagging negotiations. The Buenos Aires meeting concluded with the industrialized and emerging countries agreeing to a two-year plan of action aimed at producing binding agreements on flexibility measures, technology transfers, and limits on GHG emissions from all nations. They also committed to reach final decisions on what mechanisms would be used to achieve the emissions reductions by the end of the year 2000 (*New York Times*, November 26, 2000).

The Bush administration announced in 2001 that it would withdraw the United States from the Kyoto Protocol, and indicated it would formulate a series of voluntary programs to address the threat of climate change. In 2002, it proposed to establish a greenhouse gas registry that would permit companies making emissions reductions to register those cuts to ensure they would be recognized should a regulatory program ever be developed. It also proposed $4.6 billion in tax credits over five years as an incentive to encourage reduced GHG emissions. The Bush administration's primary approach to climate change is to reduce through voluntary measures the carbon intensity of the U.S. economy: currently, 183 metric tons of carbon dioxide are released for every $1 million in GNP; the goal is to reduce that figure to 151 tons by 2012 (Revkin, 2002).

The U.S. Congress has strongly resisted the Administration's climate change initiatives (Bryner, 2000). The U.S. Senate unanimously passed a resolution in July 1997 aimed at ensuring that the United States and other developed countries not sign a climate change agreement that did not impose on developing countries at least some (if not a similar) commitment to reduce greenhouse gas emissions (S. Res. 98, 105th Cong., 1st sess., 1997). Senate Resolution 98 specified two key conditions the Senate expected to see in any climate treaty: it "should include commitments for countries with developing economies . . . and should not result in serious harm to the economy of the United States."

As a result of Senate opposition, the Clinton administration did not submit the Kyoto agreement to the Senate for ratification. Opponents of climate change policy introduced legislation in 1999 that would prohibit the Executive Branch from implementing the Kyoto Protocol until it is ratified, and would bar any federal agency from regulating CO_2 emissions, and the provision was enacted as a rider the Department of Agriculture's FY 2000 appropriations bill.

However, in January 2003, Senators John McCain (R-AZ) and Joe Lieberman (D-CT) introduced a greenhouse gas cap and trade bill that would require sources to reduce their GHG emissions by 2010 to year 2000 levels and to 1990 levels by 2016. Companies could buy and sell credits in meeting their allowances and could meet up to 15 percent of the required emission reduction through the purchase of credits from others, including credits from carbon sequestration. Automakers could earn credits if their corporate average fuel economy standards exceed the federal fuel efficiency standard by more then 20 percent (Lazaroff, 2003). As part of the negotiations over a major energy bill that moved slowly through Congress in 2003, Senate leaders agreed to hold a vote on McCain-Lieberman in October. On October 30th, the bill was rejected by a vote of 55 to 43, with six Republicans, 36 Democrats, and Independent Senator Jeffords voting for the measure. Proponents of the bill hailed the achievement of 43 votes as a major victory, noting the dramatic shift from the 95-0 vote in 1995 in the anti-Kyoto Protocol resolution, and predicted that "the basis for a winning hand in the Senate is on the table." They predicted that while it took Congress ten years to take action on acid rain, it would take much less time to pass a climate change bill (Pianan, 2003).

By mid-2004, the Kyoto Protocol appeared to be on the verge of entering into effect. Most of the developed countries, including Japan and all fifteen European Union states, have signed the agreement, and if Russia ratifies the treaty, it will become a binding international agreement (BBC, 2002). The following kinds of trading programs are provided for under the Kyoto Protocol:

- Joint Implementation: on a project by project basis, sources in industrial countries can provide foreign investment benefits to developing countries to be used to reduce GHG emissions; these are limited to Annex I parties (industrialized and transitional economies)
- Allowance Trading: cap and trade programs that allocate allowances to sources and allow them to buy and sell allowances and require they hold adequate allowances at the end of each regulatory period to cover their actual emissions.

- The Clean Development Mechanism evolved from the idea of a clean development fund to be created by fines paid by countries that failed to meet their emission reduction goals to a program where Annex I nations can meet their emission reduction goals by investing in projects in non-Annex I nations (developing countries) that promote sustainable development.

Joint implementation and clean development mechanism projects may only count emission reductions that are additional to what otherwise might have occurred; and they must be real, measurable, and quantified against a baseline. JI projects generate emission reduction units (ERUs); CDM projects generate certified emissions reductions (CERs), also called CDM credits (Center for Sustainable Development in the Americas, 1999).

Lessons from voluntary GHG trading

A wide range of companies have developed voluntary programs to cap and reduce GHG emissions. These companies typically allow trading of emission credits across divisions and even externally to give managers flexibility in meeting emission goals. Credits can be generated by reducing emissions as well as by investing in projects that sequester carbon. British Petroleum (BP) was one of the first companies to take action on GHG emissions, pledging in 1998 that by the year 2010, BP would reduce its emissions of GHGs by ten percent from 1990 levels. To accomplish that goal, the company instituted a company-wide emissions trading system to meet its goal in the most cost effective way possible. Each business unit is allocated a fixed number of annual allowances to emit GHGs. Trading occurs across all 127 business units of BP, and units can trade CO_2 credits as well as allowances in achieving their annual emission goal. At the end of each year, if allowances are exceeded, the excess can be purchased from other units; if emissions fall below allowances, the excess may be sold. BP met that goal in October 2001, more than eight years ahead of schedule. Shell, DuPont, and many other companies have also developed voluntary programs (Bryner, 2003).

A number of U.S. cities and at least four states have developed programs to reduce GHG emissions. New Hampshire, for example requires the state's three fossil fuel power pants to cut sulfur dioxide emissions by 75 percent, nitrogen oxides by 70 percent, and CO_2 by 3 percent from 1990 levels by 2007. The plants may purchase pollution credits from out of state sources in meeting their obligations, but there are incentives for the companies to purchase them from nearby states (*Concord Monitor*, April 19, 2002). Pennsylvania and Oregon have enacted Climate Action Plans. Oregon's goal is to reduce GHG emissions by 10 percent from 1990 levels by 2010. The Climate Trust is a Portland-based NGO formed in 1997 by the law requiring power plants to reduce their GHG emissions. Plants can meet that goal by making payments to the Trust which, in turn, invests in GHG projects that avoid, displace, or sequester CO_2 emissions (Climate Trust, 2003).

Other countries are experimenting with GHG trading. Australia, Canada, France, Germany Norway, and the European Union are designing carbon trading systems, and the Danish and Dutch governments have in place pilot projects for carbon trading. The Dutch government, for example, has announced its commitment to purchase 250 million tons of credits over the next few years. The Dutch government signed a three-year $40 million contract with the World Bank to develop clean energy projects in developing countries in exchange for ten million tons of CO_2 credits. The British government launched in April 2002 the first national system for trading carbon emission anywhere in the world as part of its goal of reducing emissions by 12.5 percent by 2008. Some thirty-six companies joined the initial effort, agreeing to reduce GHG emissions by a certain amount in exchange for financial support from the government (Black, 2003). Many of these companies have found that investments in reducing carbon emissions have also reduced costs. Commitment to a GHG reduction program has prompted employees to find new ways to reduce wastes and improve efficiency.

Trading of carbon and other emissions should be pursued with caution. Emissions trading, if not carefully designed and integrated with minimum standards, can result in problems such as high concentrations of pollutants in some areas. Enthusiasm for its use may divert attention from decisions about what is required to achieve environmental goals. Trading may undermine the power of moral arguments that pollution should be reduced and conflicts with the expectation that all sources of pollution should take action to reduce emissions. It may fail to create incentives for continued technological innovation if pollution sources buy emission credits rather than invest in control equipment. Trading requires emissions monitoring capacity that may simply not exist. Some pollutants, particularly those where the total volume of pollution is the concern, are better suited for trading programs than pollutants posing localized health and environment threats. Introducing an emissions trading program, like any other policy change, creates some uncertainties and risks that regulatory officials and regulated industries may resist. Trading is not a panacea for all the difficulties posed by environmental regulation. It is not suitable for use in every regulatory program. The challenge is to determine when it should be used and, when it is appropriate, how to design and implement trading in ways that ensure that environmental protection and economic efficiency goals are achieved.

Conclusion

Emissions trading is a useful regulatory tool that can be used at all levels of government, provided that several conditions are met: the same emissions have the same effect throughout the area in which they occur; trading among different pollutants are not allowed since their health and environmental effects differ; trading of toxic chemicals, heavy metals, and other highly dangerous substances are prohibited; trading from different kinds of sources are only allowed if there is no difference in the resulting health or environmental effects; and minimum ambient concentrations are satisfied throughout the affected areas where relevant. Other constraints require emissions are quantifiable, replicable, and permanent; they are relatively easy to measure and monitoring systems are in place so that emissions can be accurately measured; a limited number of major sources that can afford the transaction costs are involved (trading schemes do not work well when there are a large number of small sources, because the transaction costs will be too high); and aggressively enforced penalties are in place for excess emissions, including fines and reductions in allowances for subsequent years. Emission limits should be ratcheted down over time to ensure environmental quality goals are achieved. The trading system should be stable and predictable. Trading should be combined with a requirement that each source make some minimum reduction in emissions, so that no source is seen as escaping at least some obligation to help solve the pollution problem. There should be clear understanding that allowances are only temporary permits to pollute; they are not permanent rights that can be changed as environmental conditions change.

Emissions trading programs can contribute to broader shifts in policy towards true cost accounting. In a political economy fundamentally committed to market exchanges, regulatory strategies that help ensure the real costs of production are included in the prices charged can make a critical contribution to achieving environmental protection goals and a sustainable economy. Economic goals of efficient use of resources are similarly fostered as prices are adjusted to reflect more accurately the real costs of production. No policy innovation will likely have a more significant impact on environmental quality and protection of natural resources than to move towards a system where the impacts on environmental factors are fully represented in prices paid. Ending subsidies and reforming price structures are the key to a more effective regulatory system, and, ultimately, more effective environmental protection. Emissions trading programs are an incremental step toward that goal. The real importance of trading programs lies in their ability to help generate support for the idea of market-based incentives, generate some data on ecological costs and benefits, and pave the way for more fundamental policy reforms that will lead to ecologically sustainable economies (Chernow & Esty, 1997).

Experience suggests that emissions trading programs work best when they are based on accurate emissions information, are built on emission limits that give adequate protection to environmental quality and natural resources, are stable and predictable, are rigorously enforced, and create more effective incentives to reduce pollution and encourage economic activities that are ecologically sustainable. Many of these conditions simply do not exist for GHGs. There is not yet in place an accurate, comprehensive monitoring system to determine CO_2 emissions and to ensure compliance with emission limits, for example. Nor is there even the basis of an adequate enforcement mechanism. There are great risks that trading will allow sources to invest in carbon sequestration projects with uncertain benefits rather than reducing their emissions.

While there is little agreement over what environmental goals should be, once they are established, market-based regulatory tools can help generate support for these proposals. The debate over environmental goals is often extremely contentious, and some of the conflict can be softened by reducing the cost of achieving those goals. The acid rain program is a classic example of how emissions trading made possible a new regulatory program with aggressive environmental goals. If economic instruments can be devised to achieve environmental goals at lower cost than conventional command and control regulation, they will become one of the most important developments in public policy and will play a major role in the move toward more ecologically sustainable societies. They can help make possible the pursuit of environmental goals that appeared too expensive to achieve with conventional regulatory approaches.

The experience of the acid rain and other trading programs in the United States suggest the importance of being able to answer affirmatively questions in the following seven areas before designing a tradable permit program:

1. Is there an accurate emissions inventory in place for determining the allocation of allowances? Will all sources be treated fairly?

2. Is an appropriate baseline used? Does it fairly reflect economic ups and downs, breakdowns and other problems with maintenance and operation, investments in and performance of pollution control equipment, and a host of other factors?
3. How are allowances allocated? Are sources that have already reduced their emissions treated fairly?
4. Are there sufficient authority and resources for effective monitoring and enforcement? Can regulatory officials ensure that quantifiable emissions reductions can be determined and that reductions are surplus, quantifiable, permanent, and enforceable?
5. Is trading among pollutants that pose different health and environmental risks prohibited?
6. Are all sources required to make at least some reductions in emissions?
7. Will environmental quality be significantly improved? Will caps on emissions be sufficient to achieve environmental quality goals?

Finally, emissions trading programs should lead to other, more powerful regulatory innovations that will more effectively encourage ecologically sustainable activities. Emissions trading programs should be designed as part of a system of emissions fees or taxes and other efforts to reflect true costs in prices and to create more powerful incentives to reduce and prevent pollution. The ultimate test of an emissions trading program is its contribution to a more fundamental shift in practices aimed at reducing pollution, improving efficiency, and conserving resources.

References

Airtrends. (1999, December 20). 1.

British Broadcasting Company (BBC). (2002). Japan ratifies Kyoto Pact. June 4, 2002, http://news.bbc.co.uk/1/hi/world/asia-pacific/2024265.stm.

Black, R. (2003). Emissions trading launches in UK. BBC News. Retrieved from http://news.bbc.co.uk/hi/english/sci/tech/newsid_1906000/1906332.stm.

Bohi, D. R., & Burtraw, D. (1997, February 4). SO_2 allowance trading: How experience and expectations measure up. Discussion paper 97, 5–6. Washington, DC: Resources for the Future.

BP. Environment. www.bp.com; "Greenhouse Gas Exchange" www.bp.com.

Bryner, G. (2000). Congress and the politics of climate change. In P. G. Harris (Ed.), *Climate change and American foreign policy* (pp. 111–130). New York: St. Martin's Press.

Bryner, G. (2003, March). Carbon trading and carbon banks: Innovations in climate change policy. Paper presented at the annual meeting of the International Studies Association, Portland Oregon.

Burtraw, D., Krupnick, A., Mansur, E., Austin, D., & Farrell, D. (1997). The costs and benefits of reducing acid rain. Washington, DC: Resources for the Future. Retrieved from www.rff.org/disc_papers/summaries/9731.htm.

Center for Sustainable Development in the Americas. (1999, June). An annotated glossary of commonly used climate change terms. Retrieved from http://www.csdanet.org/glossary.html.

Chernow, M. R., & Esty, D. C. (Eds.). (1997). *Thinking ecologically: The next generation of environmental policy.* New Haven, CT: Yale University Press.

Climate Trust. (2003). The Climate Trust Awards $1 million Contract to the City of Portland to Improve Energy Efficiency and Reduce CO_2 Emissions. Retrieved from www.climatetrust.org.

Drury, R. T., Belliveau, M. E., Kuhn, J. S., & Bansal, S. (1999). Pollution trading and environmental injustice: Los Angeles' failed experiment in air quality policy. *Duke Environmental Law and Policy Forum, 9,* 231–286.

Dudek, D. J. & Palmisano, J. (1988). Emissions trading: Why is this thoroughbred hobbled? *Columbia Journal of Environmental Law, 13,* 217–256.

Dunlap, R. E. (1992, October). Public opinion in the 1980s: Clear consensus, ambiguous commitment. *Environment, 33,* 32.

Environmental Defense Fund. (1996, September 3). U.S. Acts on Global Warming at Geneva. *Environmental Defence Fund Letter,* 1.

Environment & Energy Study Institute. (1999, January 14). Climate pact rocked by political, economic realities (pp. 12–13). Washington, DC: Author.

Environment NOW News Service, (2000). NO_X Emission Trading Credits Reach Record High. Retrieved April 28, 2000, from http://www.environ... display/misc.asp?tab=market&id=180.

Environmental Reporter. (1992, May 29). California: Problems seen setting baseline levels in south coast emissions-trading program, p. 437.

Flavin, C. (1995, July–August). Climate policy: Showdown in Berlin. *World Watch, 8,* 8–9.

Flavin, C. (1998, November–December). Last tango in Buenos Aires. *World Watch, 11,* 10–18.

Gartner, M. (1992). A skeptic speaks. *Environmental Protection Agency Journal, 18*(2), 26.

Hahn, R. W. (1988, Winter). Innovative approaches for revising the clean air act. *Natural Resources Journal, 28,* 171–184.

Hernandez, R. (2000, May 2). Albany battles acid ran fed by other states. *New York Times.* A1.

Holtcamp, J. A. (2000, January). Emissions trading on local, regional, national, and international scales. Presentation at the Rocky Mountain Mineral Law Foundation conference, Air Quality Regulation for the Natural Resources Industry, Salt Lake City, Utah.

Kelman, S. (1981). *What Price Incentives?* Boston: Auburn House.

Korb, B. (1998). U.S. experience with economic incentives for emissions trading. In Organisation for Economic Cooperation and Development.

Applying market-based instruments to environmental policies in China and OECD countries (pp. 99–115). Paris: Author.

Kosloff, L. H., & Trexler, M. C. (1997, January). Global warming, climate-change mitigation, and the birth of a regulatory regime. *Environmental Law Report, 27*, 10012–10018.

Lazaroff, C. (2003, January 8). U.S. Senate bill would cap greenhouse gas emissions. *Environmental News Service.* Retrieved January 9, 2003, from http://ens-news.com/ens/jan2003/2003-01-08-06.asp.

Lee, Jennifer S. *2 Senators Aim to Put Others on Record on Emission Cap, New York Times,* July 28, 2003, www.nytimes.com/2003/07/28/politics/28ENV1.html. Accessed July 29, 2003.

Natural Resources Defense Council. (1996, May). *Breath taking: Premature mortality due to particulate air pollution in 238 American cities.* New York: Author.

Oberthuer, S., & Ott, H. E. (1999). *The Kyoto Protocol: International climate policy for the 21st century.* Berlin: Springer.

Pianan, E. (2003, October 21). On greenhouse gas senate rejects mandatory cap emissions, *Washington Post.* A4.

Rachel's Environment & Health Weekly #628. (1998, December 10). Sustainable Development, Part 5: Emissions Trading. Available through rachel@rachel.org.

Repetto, R., Dower, R.C., Jenkins, R., & Geoghegan, J. (1992). *Green fees: How a tax shift can work for the environment and the economy.* Washington, DC: World Resources Institute.

Revkin, A. (2002, February 14). Bush plan expected to slow, not halt, gas emission rise. *New York Times.* A1.

Seligman, D. A. (1994). Air pollution emissions trading: Opportunity or scam? Unpublished paper. Washington, DC: Sierra Club.

South Coast Air Quality Management District. (1996, January 12). *AQMD News.* Retrieved June, 1999, from http://www.aqmd.gov/news1/Archives/recaudit.html.

Stavins, R. (1989). Harnessing market forces to protect the environment. *Environment, 31*(1), 5–35.

Stevens, R. (1997, August 29). California pollution officials suspend emissions trading. *Wall Street Journal.* A4.

The great Berlin greenhouse compromise. (1995, April 27). *Nature, 374*, 749–750.

U.S. Congress, Office of Technology Assessment. (1995). *Environmental policy tools: A user's guide.* Washington, DC: Government Printing Office.

U.S. Council on Environmental Quality. (1995). *Environmental Quality: Twenty-fifth Anniversary Report.* Washington, DC: Council on Environmental Quality.

U.S. Environmental Protection Agency. (1999). Market Incentives Resource Center. Retrieved June, 1999, from http://www.epa.gov/oms/transp/traqmkti.htm.

U.S. General Accounting Office. (1997, July 9). Overview and issues on emissions allowance trading programs. GAO/T-RCED-97-183.

U.S. General Accounting Office. (2000, March 9). Acid rain: emissions trends and effects in the eastern United States. RCED-00-47.

U.S. National Science and Technology Council. (1998, May). NAPA biennial report to congress: An integrated assessment.

Wirth, T. E., & Heinz, J. (1988, December). Project 88, Harnessing Market Forces To Protect Our Environment. Unpublished report.

Wirth, T. E., & Heinz, J. (1991, May). Project 88—Round II, incentives for action: Designing market-based environmental strategies. Unpublished report.

CHAPTER NINE

Environmental Impact Statements

Gift Box or Black Box?

WALTER A. ROSENBAUM

In 1995, on the twenty-fifth anniversary of the National Environmental Policy Act (NEPA), the Environmental Law Institute (ELI), one of Washington's most influential environmental think tanks, launched a fiesty broadside against the President's Council on Environmental Quality (CEQ). The CEQ is NEPA's statutory guardian, it's most important federal advocate and administrative interpreter. Not coincidentally, CEQ was celebrating NEPA's first quarter century with literature largely advertising NEPA's achievements. In ELI's opinion, however, the CEQ hadn't yet gotten NEPA right. Dismissing much about CEQ's understanding of NEPA as sheerest nonsense, they administered a thorough scolding:

> The conventional wisdom about NEPA is that it is a flowery Preamble attached to a single–and wholly procedural—requirement to prepare environmental impact statements for a small subset of federal decisions. This conventional wisdom is wrong. . . . It is, rather a vision for this nation's future, coupled with an intensely practical strategy for action. . . . If we want to solve our problems as a nation and pursue a path toward sustainability, we have the law we need for effective action—unused in its original box. (Environmental Law Institute, 1995, pp. 1–2)

Thus, a public brawl among the environmental community and perversely appropriate for NEPA's twenty-fifth Anniversary. From its inception, NEPA—only three pages long—has been a matter of continuing legal and political contention about its purpose and implementation. Here, amid fractious discourse, begins the search for the impact of federal Environmental Impact Statements (EISs), NEPA's most well-known statutory creation.

The contention about intention

EISs, a mandated federal administrative procedure since NEPA's enactment in 1970, have generated a massive body of evaluative literature (Cohen, 1980; Lazear, 1978; Hill & Ortolano, 1978; Adreen, 1990; Blaug, 1993; Belsky, 1998). Whether this literature testifies to the effectiveness of the EIS process—and, if so, to what extent—depends upon what one assumes the process was supposed to accomplish. NEPA at its inception was widely acclaimed by environmentalists as a Magna Carta for environmental protection, but its Congressional intent, and much of its brief statutory language, have been disputed since its inception. The crucial issue is whether the statutory language defining NEPA's purpose in Section 101 is the engine driving the EIS process and setting its direction—in short, whether the sections of NEPA defining its purpose and prescribing the character of federal EISs are inextricably related.

A contested Magna Carta

Many environmental advocates, scholars and legal experts, believe the EIS process has been gravely flawed because Congress, the federal courts, and the White House have misconstrued or ignored the statutory intent of NEPA, thereby weakening the scope and impact of the EIS process. In effect, NEPA has been politically and judicially hijacked. Like Lynton Caldwell, one of NEPA's primary draftsmen and most outspoken academic advocates, they contend that NEPA's goals and principles have from their inception "been treated as noble rhetoric, having little practical significance" (Caldwell, 1998, p. 205; see also Lindstrom, 2000; Bartlett, 2000). In a similar vein, the ELI speaks for many among the environmental legal community in deploring the crabbed view and neglect of NEPA's actual intent commonly found among federal bureaucrats, judges and White House administrators (Environmental Law Institute, 1995, Chapter IV).

A RIGOROUS AND VISIONARY NEPA

In the view of NEPA's most aggressive partisans, NEPA's Section 101 is nothing less than a substantive policy command. It is a visionary declaration of positive law, mandatory upon Congress, the President, the courts and all federal agencies, to protect environmental quality and promote ecological productivity in six specific ways:

Concern for the future: "fulfill the responsibilities of each generation as trustee of the environment for succeeding generations."

Environmental equity: "assure for all Americans safe, healthful, productive and esthetically and culturally pleasing surroundings.

Beneficial use: "attain the widest range of beneficial uses of the environment without degradation, risk to health or safety, or other undesirable and unintended consequences."

Historical, cultural, and biological diversity and individual liberty: "preserve important historic, cultural, and natural aspects of our national heritage, and maintain, whenever possible, an environment which supports diversity and variety of individual choice."

Widespread prosperity: "achieve a balance between population and resource use which will permit high standards of living and a wide sharing of life's amenities."

Management for quality and conservation: "enhance the quality of renewable resources and approach the maximum attainable recycling of depletable resources" (45 U.S.C. Para. 331(b)(1)-(6)).

In this perspective, the six objectives are a template routinely applied to any federal actions significantly affecting the human or natural environment. Armed by this rigorous policy mandate, NEPA's Section 102 with its EIS process would provide federal decision makers with scientific and other relevant technical information driving policy decisions in the direction of these six environmental objectives.

This tough-minded approach is disputed by other legal scholars and policy experts. They assert that NEPA's Section 101 was intended to promote, not to mandate, federal agency decisions consistent with its objectives and that the EIS process in Section 102 was intended to provide the technical information to encourage such environmentally protective policymaking. (Taylor, 1984; Dennis, 1997; Mandelker, 1984). Proponents of this more relaxed interpretation have clearly won the battle in the federal courts and agencies but legislative history and closely related documents support the tougher view of NEPA's intent (Lindstrom & Smith, 2001; Peterson, 1971; Andrews, 1976; Caldwell, 1997). A succinct expression of this rigorous intent is found in the Senate Committee on Interior and Insular Affairs' report in 1969 defining NEPA's purpose at the time of its enactment:

> The policies and goals set forth in Section 101 can be implemented if they are incorporated into the ongoing activities of the Federal government in carrying out its responsibilities to the public. In many areas of Federal action there is no body of experience or precedent for substantial and consistent consideration of environmental factors in decision making . . . To remedy present shortcomings in the legislative foundation . . . pro-action-forcing procedures will help to insure that the policies enunciated in Section 101 are implemented. (U.S. Senate, 1969; see also Bear, 1995)

In this perspective, the EIS process was an intentional marriage of substantive policy mandate with procedural requirements for its implementation (Lindstrom & Smith, 2001, p. 62). NEPA's policy mandate remains largely a memorial to failed Congressional intentions and, consequently, the EIS process has never acquired the legal and political impact it was meant to possess.

What went wrong? The short answer is that the miscarriage of NEPA's substantive intent is the consequence of a feeble political will in Congress and the White House, compounded by a severely restrictive judicial interpretation of NEPA and its impact statement requirements (Fogelman, 1990; Plater, Adams & Goldfarb, 1992; Swartz, 1995). In the three decades since NEPA's enactment, Congress had demonstrated scant interest in reasserting NEPA's substantive intentions or fortifying the EIS process politically or procedurally. In fact, only five NEPA oversight hearings have been held since 1988 (U.S. Senate, 1991; U.S. Senate, 1992; U.S. Senate, 1995; U.S. Senate 1996, U.S. House of Representatives, 1995). Nor has Congress been inclined to recognize any special need for added resources among the agencies required to perform EIS analysis. And, as subsequent discussion will illuminate, the federal courts have not wavered in their restrictive interpretation of NEPA's intent. Nonetheless, the EIS process, even when divorced from NEPA's contested substantive intent, has had significant political, scientific, and environmental impacts reaching beyond the specific federal agency actions to which it is addressed and affecting resource politics in some unanticipated ways.

The anatomy of the EIS

The EIS process itself is described in a single concise paragraph of NEPA. Section 101(c) requires that all agencies of the federal government shall include in every recommendation or report on proposals for legislation and all other major Federal actions significantly affecting the quality of the human environment, a detailed statement by the responsible officials on:

(i) the environmental impact of the proposed action,

(ii) any adverse environmental effects which cannot be avoided should the proposal be implemented,

(iii) alternatives to the proposed action,

(iv) the relationship between local short-term uses of man's environment and the maintenance and enhancement of long-term productivity, and

(iv) any irreversible commitments of resources which would be involved in the proposed action should it be implemented. (42 U.S.C. Para. 4332)

The CEQ provides detailed regulations to guide agencies in the preparation and review of EISs and the U.S Environmental Protection Agency (EPA) is required by the Clean Air Act to review all EISs. Within a few years of NEPA's enactment, most federal agencies likely to prepare a significant number of EISs had hired specialized staff, or created designated offices, for EIS preparation. As always, a flourishing subeconomy of consulting firms also materialized and multiplied to assist agencies in this task. By the beginning of the 1980s, national organizations representing the rapidly developing environmental assessment professions, also appeared. In short, within a decade of its creation, EIS preparation had become professionalized and bureaucratized.

The process: Uncoupled but still important

The courts—with tacit congressional approval—may have uncoupled the EIS process from NEPA's statutory mission, but the EIS process has nonetheless changed the style of federal resource policymaking in significant, and sometimes unintended ways. Perhaps the most durable impact is that the EIS process has created a new structural element embedded with the broader institutional configuration of federal resource policymaking. This has significantly reordered the relative influence and styles of decision making between the traditional participants. It has, at least incrementally, altered the federalized structure of national resource management. These repercussions have as much to do with the political controversy attending the EIS process as any other aspect of its impact. The larger lesson seems more apparent now than in the 1970s: any effort to empower science and related disciplines in resource decision making through procedural change involves, however unintentionally, a disruption and reordering of political and institutional influence.

Another significant consequence of the EIS process is its impact upon the development and utilization of environmental science. The EIS procedure was intended to promote a new integration of science with environmental policymaking through the use of interdisciplinary information. The actual extent of this integration is contested but the EIS procedure (as its creators intended) created a vigorously enlarging data base of environmental information, much of it publicly available. The impact on water resource science, for example, is pervasive.

> "NEPA permanently changed the landscape for engineers, scientists, and everyone else involved in water resource issues—by altering the character and scope of acceptable professional practices," noted a 1993 National Academy of Science Report on water resource science. "Once the compelling character [of NEPA] sunk in, it became clear to everyone in the water resources community that the fundamental policy issues targeted by NEPA would require new perspectives, insights, and skills" (Lynn, 1993, pp. 59–60. see also National Academy of Science, 1995, chap. 2–4).

NEPA has created new, or enhanced, federal funding for environmental sciences and related disciplines. It has encouraged the growth of epistemic communities specialized in environmental impact assessment, including the collection and integration of interdisciplinary information into environmentally-relevant formats. In effect, it has stimulated the construction of new knowledge growing from environmental imperatives. EIS preparation also changed the scientific base for decisions about nuclear power plant siting and promoted the growth of risk assessment research at major federal laboratories (ORNL, 2002). This ripple effect of the EIS process is, perhaps, its most enduring ecological legacy and certainly its most consequential scientific impact.

In many instances, the EIS process is an early warning system for environmental advocacy and science communities, even though public involvement in the process is often more restrictive than its authors intended. The statutory and regulatory requirements for early public disclosure and review of agency EIS statements often provide interested and affected parties with a translation of agency programs into comprehensible and relevant language which, in turn, incites political mobilization. This public disclosure is valuable not only to the

Washington-based national organizations (that sometimes employ specialized staff for EIS oversight) but especially to smaller state and local organizations otherwise lacking the resources to acquire and interpret the complex and often (deliberately) mystifying, bureaucratic, syntax-adorning program descriptions.

Finally, the EIS process promotes more inter- and intra-agency collaboration in program planning and assessment and a greater concurrent sensitivity to the environmental impacts of program development than would otherwise have occurred. Agency decision makers routinely consider a relevant EIS, at least to the satisfaction of the federal bench, when decisions within the jurisdiction of NEPA are contemplated. The critical issue, however, is the environmental consequences of the process. Evidence about the scale and character of these impacts is impressionistic, and empirical measures, such as the number of EISs prepared or surveys of EIS litigation, often amount to administrative bean counting rather than reliable environmental information.

More sophisticated efforts to demonstrate NEPA's beneficial impact on agency policymaking have been made in several more or less comprehensive surveys of agency practices and in the collection of numerous case studies or surveys of presumably knowledgeable experts (Council on Environmental Quality, 1998; U.S. Department of Transportation, 2003; National BLM Wilderness Campaign, 2003; Smythe & Isber, 2003; Frye, Yates, & Batts, 2001; Tzoumous & Finegold, 2000; Canter & Clark, 1977; Bear, 1993; Blaug, 1993). The available evidence suggests EISs are more likely to compel incremental, though sometimes environmentally valuable, modifications in major federal programs rather than to halt such programs affecting the environment. Unfortunately, efforts of agencies to mitigate the adverse environmental impact of their programs in response to EIS-related information are seldom monitored after program implementation.

The number of formal EISs has steadily declined in the decades since 1970 (see Figure 9.1), largely because agencies have substituted the shorter, less rigorous Environmental Assessment (EA) which allegedly has resulted in an important—but not empirically measurable—"reduction in adverse [program] effects below significant levels" (CEQ, 1997, p. 3). An EA is, according to the CEQ, a "concise public document . . . that serves to . . . briefly provide sufficient evidence and analysis for determining whether to prepare an environmental impact statement or a finding of no significant impact" (CEQ, Regulations, Sec. 1508.9). Most often, EAs are used to demonstrate that an agency action will not have a significant impact on the environment sufficient to require a formal EIS or that any significant impact will be mitigated by measures associated with the action and which will be identified in

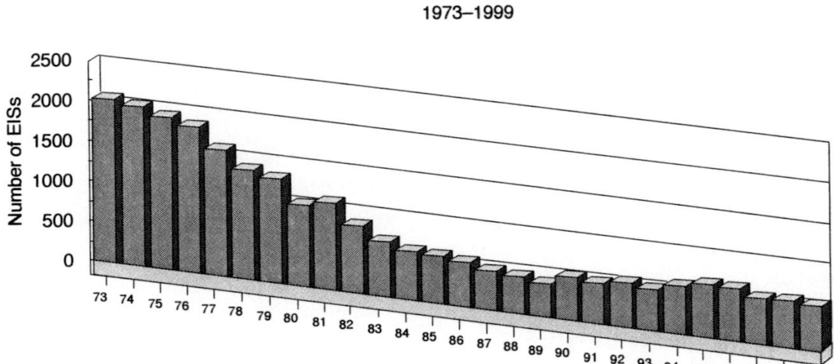

Figure 9.1

a Mitigated Finding of No Significant Impact or FONSI. Thus, the Forest Service might issue an EA to demonstrate that a proposed road through a federal forest will not have a significant impact on the environment or that mitigation measures to stabilize the soil during road construction will be created and identified in a FONSI.

The impact of the EIS process on federal policymaking has been diminished by both Congress and the White House when political expediency and opportunity were compelling. Indeed, political intervention has been an important reason for federal circumvention of the NEPA process over the last several decades. This political intervention assumes several forms. For instance, legislators have increasingly used appropriations riders in the last decade to overturn federal public lands policies negotiated through the NEPA process among federal agencies, private interests, and state governments (Zellmer, 1997). President Clinton, in turn, made numerous public land decisions, including National Park and National Monument designations, during his terms in office under authority of the American Antiquities Act (1906), which he asserted exempted the agencies with jurisdiction over such land from requirements of the NEPA process. Additionally, all federal agencies have rejected the notion—with consistent presidential approval—that the NEPA process should apply to their overseas actions as well as their domestic decisions. Thus, Department of Defense military installations pollute soil and groundwater throughout Europe without any NEPA review and, elsewhere abroad, "other federal agencies support actions such as hydroelectric development, mining, and pesticide spraying without informing the public—or learning themselves—of the environmental consequences of such action" (Manheim, 1994). Such political intervention in the NEPA process, while sporadic, has been appealing because it makes judicial review of federal decisions affecting the environment less likely.

Despite its vulnerability to political intervention, the array of impacts associated with the NEPA process are still significant, sometimes unexpected and undervalued, but overall deserving of attention.

A new decision making infrastructure

Formally, the EIS process mandates an overt and legally grounded structure of EIS creation, circulation, review and oversight within a larger institutional design of federal resource policymaking. This formal EIS structure embraces the CEQ and the bureaucratic units responsible for EIS production; it requires procedural compliance within the affected agencies and consideration of the resulting EISs by the EPA's Office of Federal Activities which reviews all EISs, by the federal courts exercising oversight of the process, and by the Executive Office of the President. In actuality, the federal courts are the most salient among these institutions in determining the character and impact of the EIS process. That influence, in turn, affects the relationship between the states and federal government in resource decision making and, especially, in public lands management.

The courts and NEPA

NEPA was judicialized from its inception, and thereby captured within an institutional style inherent to American conflict management. In a political culture where litigation legitimates and resolves policy conflict, federal judges are policymakers, the judicial process a philosophical template in shaping the consequences of legislation. In the case of NEPA, this judicial intervention shapes the impact of the EIS process in several fundamental ways.

Most significantly, a continuing series of federal court decisions, beginning with *Calvert Cliffs' Coordinating Committee v. Atomic Energy Commission* (1971), have consistently held that Section 101 of NEPA is flexible and "leaves room for a responsible exercise of discretion and may not require particular substantive results in particular problematic instances"—in essence, that NEPA embodies not a set of six substantive, and enforceable policy mandates but a statement of desirable policy goals, not obligations, which should inform federal agency decisions (*Flint Ridge Development Co. v. Scenic Rivers Association of Oklahoma,* 426

U.S. 776 [1976]) is an equally seminal, if later, interpretation). While agencies are obligated by NEPA to create and consider environmental impact statements when relevant to their policy decisions, they are not compelled to make decisions compatible with the content or conclusions of an EIS. The practical implication is that impact statements are a procedural requirement to which agencies must demonstrate good faith compliance. But federal agencies are not obligated to use the impact statements as a basis for making decisions that promote the policy goals stated in NEPA's section 101. Thus, it is virtually impossible to overturn substantive agency decisions on grounds of incompatibility with the environmental purposes of NEPA. But federal judges have no aversion about requiring agencies to comply rigorously with the EIS procedural requirements or meticulously scrutinizing the technical and scientific documentation to determine its statutory acceptability. These consequences shape the impact of EISs in many important ways.

By implicating themselves so pervasively in the EIS process, federal judges have become the most authoritative delineators of the EIS procedural design and the final arbiters of the scientific quality of EIS statements. In both respects, this has major consequences for the impact of the EIS upon federal government decision making. EISs are now commonly thick, dense, and exceedingly technical documents—the average completed EIS is about 580 pages of which about two-thirds is non-text—usually bloated by elaborate and excessive scientific, technical and administrative detail to bulletproof the document from judicial findings of procedural or documentary deficiency (U.S. Environmental Protection Agency, Office of Federal Activities, 2003). (The fifty pound EIS prepared for the Trans-Alaskan Oil Pipeline has already passed into legend). Often, this obscures the relevance of technical data and the clarity of scientific inference in a melange of ill-assorted detail. The CEQ has found this a continuing problem:

> In addition to gathering more and better data, NEPA practitioners need to analyze existing information more effectively. EISs too often have more data than required to make a responsible decision, but not enough analysis of the data focuses on the decision. What is often lacking in EISs is not raw data but meaning. (U.S. Environmental Protection Agency, Office of Federal Activities, 2003: "EIS Page Lengths")

In an effort to avoid this bulletproofing, agencies have increasingly substituted the shorter, and far less technically or scientifically rigorous 'Environmental Assessments' (EAs) often leading to a FONSI ('Finding of No Significant Environmental Impact). An EA is "a concise public document that analyzes the environmental impacts of a proposed federal action and provides sufficient evidence to determine the level of significance of the impacts." A FONSI presents "the reasons why any action . . . will not have a significant effect on the human environment and for which an environmental impact statement therefore will not be prepared (Council on Environmental Quality, Regulations, Sec. 1508.13). In both instances, agencies essentially produce an analysis demonstrating that any anticipated adverse effects of their actions are below a level significant enough to require a formal EIS. Agencies commonly assert these mitigation efforts achieve the desired environmentally protective consequences but often do not subsequently monitor the mitigations to determine actual impacts. It is also argued that proliferation of EAs and FONSIs diminishes the quality of scientific information available to both the public and agency officials responsible for the affected decisions.

Judicial oversight of the EIS process, with attendant impacts upon agencies preparing EISs and the courts reviewing them, is inherently self-perpetuating: judicial review and adjudication of EISs begets further litigation and bureaucratic improvisions and so on. Judicial involvement is often asserted to increase (often severely), the time, cost, and frequency of agency EIS preparation. In fact, the number of EIS has diminished dramatically since NEPA's inception. Data concerning the average cost and duration involved in EIS preparation is elusive but available studies suggest both have been inflated to an indeterminate but significant extent by judicial intervention in the process. However, the substantial decline in EISs since the early 1970s betrays a less suspected judicial impact on EISs procedures. Vigorous judicial oversight of formal EISs review is one incentive for agencies increasingly to substitute an EA for the more demanding EIS in many instances where an EIS would also be acceptable.

Currently, federal agencies produce more than 50,000 EAs annually. An EA substantively and procedurally simplifies the EIS process while still providing considerable information relevant to the environmental consequences of agency decisions. The CEQ issued regulations in 1972 sanctioning and defining essential elements in the EIS process (CEQ, Regulations, Sections 1501.3, 1501.4). The federal courts generally reject challenges to agency use of EAs and FONSIs in determining that a proposed action does not meet NEPA's significance test, or in demonstrating that agency activity will mitigate any significant impacts of agency decision, so long as the EA has a reasonable basis (See *Hanley v. Kleindienst,* 471 F. 2nd 823 (2nd Cir 1972), *cert. denied,* 412 U.S. 908 (1973); and *Friends of Fiery Gizzard v. Farmers Home Administration,* 61 F. 3rd 501 (6th Cir 1995)). However, the courts have not hesitated on occasion to scrutinize closely the documentary basis of such determinations. CEQ officials, additionally, have frequently commended the EA as a time- and money-saving alternative to the EIS (See, e.g., testimony of CEQ Chairman Kathleen McGinty in U.S. Senate, 1996). However, opportunities for public engagement, scientific exploration of proposed agency actions, and exploration of related issues are materially diminished so the EA is, in important aspects, a very different document than an EIS qualitatively. To the extent EAs degrade more rigorous standards required of EISs, critics often consider them to be evasions of NEPA's intent and congenial to bureaucratic decisions with ultimately undesirable environmental consequences.

EISs and federalism

State resource officials, particularly Governors and extractive industries in the Western states (which embrace most of the vast public land administered by federal agencies), frequently assert that judicial oversight of federal resource agencies has severely diminished the role of state governments in creating and assessing EISs, thereby adversely affecting the balance of powers in a federalized structure of land management. It is the courts, not the states, to which federal land management agencies seem most deferential when decisions affecting local public lands are an issue, according to the complaint. Insofar as EISs do influence agency decisions, this amounts to a kind of back door judicial influence on agency decision making. "It's not just that the courts are directly involved in managing many of our resources," complained the Attorney General of Wyoming to the 1998 congressional committee, "they are indirectly managing all of them in our state because of the fear of litigation, not just because of actual litigation" (U.S. House of Representatives, 1998: 44). Significantly, when Congress did conduct a rare oversight hearing of NEPA in 1998, virtually all the submitted statements critical of NEPA impact (and most of the witnesses who appeared) represented states west of the Mississippi River.

Federal restrictions upon resource development in western federal lands is an historic grievance which doubtless animates some of this backlash against the EIS process and the federal courts allegedly involved in its abuse. Of more relevance is that these western spokesmen commonly cite the quality of science prescribed in EIS regulatory codes as a major reason for their diminished influence in federal land management decisions. In such a manner, the contested construction of science in EIS design has become a surrogate for conflict between federal and state governments over allocation of public land resources—in effect, a power struggle by proxy. These proxy disputes focus upon the alleged inattention to socio-economic impacts in agency EISs and EAs and the need for more 'adaptive management' in EIS-inspired environmental mitigation plans often required of extractive permits on federal lands.

Federal conflict by proxy: EIS science

Criticism that EISs and EAs give inadequate attention to the socioeconomic impacts involved in federal actions most often has a western inflection. Western politicians and the extractive industries seeking more expansive access to western public land resources have been the most outspoken of these critics, and it is the economic element that most concerns them. Typically, complaints about the lack of socioeconomic analysis arise in a situation where a federal agency—for instance, the Department of the Interior—restricts access to federal resources because an EIS or an EA anticipates significant adverse ecological consequences or harm to an endangered species—as often happens when the Bureau of Land Management designates large tracts of federal timberland 'roadless areas' and therefore inaccessible for commercial timbering. Characteristically, the argument is made that the ignored economic impacts include income lost to business or communities from the prohibition on resource extraction. Here, for instance, criticism concerns restricted access to federal grazing lands imposed upon the cattle industry. The speaker is Michael J. Byrne, an official of the National Cattlemen's Beef Association:

> the way NEPA is currently being administered is subverting the whole purpose of the Act . . . Congress stated that it is the policy of the Federal Government to create and maintain conditions under which man, and I underscore man, and nature can coexist and maintain productive harmony and fulfill the social, economic and other requirements of present and future generations of America. Instead, NEPA has evolved . . . into a unbridled regulatory apparatus which subordinates the economic needs of the community to agency preferences for resource preservation. This situation causes uncertainty and apprehension in the ranching community. (U.S. House of Representatives, 1998: 48)

What of NEPA's language? NEPA implies social and economic impacts of proposed federal actions should be considered in the EIS process but in what manner, at what stage, and to what extent is unclear. Subsequent CEQ regulations and litigation have established that socioeconomic analysis, including the kind of considerations the critics demand, must be included in an EIS but not necessarily in the EAs now commonly substituted for an EIS. It is this situation which prompted Kathleen McGinty, Chair of the CEQ in the Clinton Administration, to agree with much of this criticism (U.S. House of Representatives, 1998, p. 50). No action was taken by the Clinton Administration, however, to revise CEQ regulations requiring a more vigorous socioeconomic analysis in the rapidly multiplying EAs. One might expect a more aggressive response from the George W. Bush administration, whose sympathies are clearly on the side of the critics.

Adaptive management is a more recent interpolation into the debate over the construction of EIS science. In the last decade, the term has increasingly appeared in discussions of innovative resource management and sustainable development. Although still subject to variable interpretation, the National Academy of Science provides a reasonable generic definition: "adaptive management employs scientific monitoring and research to measure and explain the effects of management actions. Results of monitoring and research are then used to adjust future management strategies" (National Research Council, 2000, p. 2). This is in contrast to more comprehensive, inflexible resource plans, whose objectives and techniques, once formulated, are implemented with relatively little modification. Adaptive management techniques are now sometimes advocated, or implemented in resource management strategies associated with EISs and EAs, primarily as an approach to mitigating the undesirable effects of some proposed federal action. However, the political and corporate exponents of more adaptive management in federal EISs or EAs are likely to interpret the term with very different implications. "Make a decision based upon the best information at the time, don't try to cover every possible contingency," explains the governor of Wyoming. "You can always ask one more question that starts off with 'what if .' " Make the decision, get underway and monitor the performance and if there is impact, adapt to correct the problem"—essentially the same definition as CEQ's McGinty (U.S. House, 1998, p. 50).

This freewheeling definition resonates well with those who believe the EIS and related procedures involves too much data gathering, too great a concern for technical and scientific accuracy in long-term planning. Indeed, many scientists and resource planners quite removed from the EIS battlefront might agree to this interpretation. Not coincidentally, however, this relaxed definition appeals to those explicitly or implicitly desiring more opportunity to influence the implementation of mitigation plans, to interests who want scientifically defensible grounds for intervention in the EIS process for political or economic reasons. Thus, adaptive management may be simultaneously viewed as an exercise in resource management science or political science, an item of contested science whose eventual construction and application within the EIS and EA process will have material political and economic consequences for the balance of political forces in the federal structure of resource planning.

The EIS and environmental science

Almost all the appraisals of the EIS process, popular or academic, have been produced by scholars or organizations associated with the legal profession or policy advocacy not by the science community. Yet in many respects the most obvious, and some might well argue the most enduring, impact of the EIS process has been on the quality and use of ecological information, and not in governmental policymaking alone. This was at least one clear and successful intention of NEPA's authors. The EIS procedure was intended to force "a restructuring of the uses of information—notably scientific information—in the process of ongoing planning and decision-making" because "without this strategy there was nothing to compel the agencies to give more than token recognition to the purposes and provisions of NEPA" (Caldwell, 1982, p. 10).Whether it has forced such consideration is frequently problematic. Whether it has promoted the creation and proliferation of such information is certain.

The need for scientific data to support NEPA's mandate that an EIS should, among other things, identify any adverse environmental effects which cannot be avoided should the proposal be implemented, and explore the relationship between local short-term uses of man's environment and the maintenance and enhancement of long-term productivity,—requirements which could not be construed as discretionary by the courts—created a significant demand for scientific and technical information that had hardly existed before 1970 or existed not at all. This included such information as the character and extent of natural ecological cycles, the definition of carrying capacity, and the biological implications of biodiversity as well as a great many other matters. The EIS process, in this respect, reinforced the demand for such information also created by many other aspects of environmentalism (Hays, 1987, p. 8). The demand for new scientific information to conform to agency EIS mandates (a need abetted by judicially imposed requirements), accelerated the growth of ecology as a scientific discipline and ecologists as an epistemic community. The impact statements themselves are indexed according to project, negative and positive impacts, and legal mandates. The resulting document, *EIS: Digests of Environmental Impact Statements,* is now a standard reference concerning the environmental consequences from a wide variety of human activities.

Clear evidence of the EIS impact upon environmental science has been the strong impetus it creates for agencies with high demand for EIS data (the Department of the Interior, the Army Corps of Engineers, and the National Forest Service) to develop environmental indicators, to acquire better baseline information on environmental trends over geography and time, and to utilize new technologies, such as the Geographic Information System (GIS), to interpret ecological data. The EIS has also been a major reason for the relatively recent growth of empirical and analytic techniques that attempt to assess the cumulative environmental impact of federal policies and, in a broader perspective, to develop a science of cumulative impact measurement. Cumulative impact is considered to be "the impact on the environment which results from the incremental impact of an action, when added to other past, present and reasonably foreseeable actions regardless of what agency, federal or nonfederal, or what person undertakes the action" (CEQ, 1997, p. 4). One recent consequence of this focus on cumulative impact has been the development of the first systematic protocols for cumulative impact assessment, including one created by the CEQ itself, which serve a scientific constituency much broader than federal agencies and address significant problems of general scientific concern (CEQ, 1997).

The need for high quality information in EIS documents has also stimulated the development and utilization of new information technologies to improve NEPA decision making. These, and the data bases to which they are related, have frequently become widely available to the public and the scientific community, especially with the rapid evolution of the Internet as a national and international repository of ecological information. This new information technology involves websites providing access to information about the preparation, use, and resources available for environmental impact assessment, such as the CEQ's own NEPAnet (http://ceq.eh.doe.gov/nepa/nepanet.htm). A number of agency websites now provide a great variety of ecological data relevant not only to EIS preparation but to a wide diversity of scientific disciplines. The U.S. Geological Service's (USGS) "Environmental Impact Analysis Data Links," for instance, provides users with access to data bases involving domestic agriculture, endangered species, energy, hydrology, wetlands, and much else (http://h2o.usgs/.gov/public/eap/env_data.html). The Census Bureau offers a website including a simple mapping system that relates population, income and ethnicity data with environmental data from the EPA, the Nuclear Regulatory Commission (NRC), the USGS and numerous other agencies. Currently, the growing edge of this technological innovation includes an increasing number of agency websites providing models, or information about models, which may be used to integrate environmental data with other information (many of these technological innovations are identified in CEQ, 1998, Part I; see also Jesse, 1998).

The enrichment and diversification of environmental data and related technologies has been accompanied by the growth of professional associations representing environmental assessment specialists and the expansion of older scientific organizations to incorporate environmental assessment within their professional focus. The International Association for Impact Assessment illustrates the rapid internationalization of the environmental assessment profession within the last decade and the evolution of an epistemic community networked across national and regional boundaries. The U.S.-based National Association of Environmental Professionals illustrates another variant of professionalization in impact assessment where EIS specialties are embedded within a more comprehensive vocational cluster of ecological professions, many of them involved in commercial consultation.

Proponents of NEPA assert that federal legislation has been a major reason for the rapid growth of numerous national environmental impact assessment organizations, international publications, and a slowly evolving network of regional and global assessment institutions. The internationalization of environmental impact assessment is an undoubted reality. Most major national governments, including all European and major Asian states, today claim to have some operational variant of an EIS process associated with various policy mandates for national environmental conservation. The internationalization of governmental environmental impact assessment, often in conjunction with environmental NGOs, has been most successful in expanding ecological data bases, creating new ones, and encouraging the growth of ecological science. The impact of this data, and the concept of an environmental assessment process, upon national policymaking is less obvious and highly variable between governments.

While the demand for information generated by federal agencies, or their surrogates, to document NEPA-related actions has undoubtedly enriched the scientific base for ecological decision making in the federal government, a persistent deficiency has been long-term data created by monitoring the impact of federal actions after NEPA-related decisions have been made. CEQ regulations and related litigation still do not require such monitoring although some agencies, such as the Department of Defense and the Department of Energy, now include monitoring in their own NEPA regulations. The need for monitoring data would undoubtedly increase to the extent that adaptive management is adopted as an agency strategy for implementing their NEPA-related planning.

An early warning system

Among the many predicted benefits of NEPA, promoters commonly cited opportunities the EIS process would provide for greater public involvement in agency decision making and for broader dissemination of public information about agency decisions affecting the environment. In effect, the EIS process was conceived as a public tripwire, alerting affected stakeholders to the ecological implications of decisions about which they might otherwise be ignorant and presumably inciting environmental organizations to mobilize when they perceived themselves affected adversely. The subsequent CEQ guidelines for agency EIS preparation requires public notice and opportunities for public comment concerning both draft and final EISs and requires agencies to provide the public with reasonable assistance in obtaining copies of both.

These prospects aroused critics to predict a multitude of objectionable consequences: long and costly delays in constructing needed public works, a proliferation of harassing litigation causing more delay and expense, an enlargement of environmentalist political power at the expense of extractive industries using public land, a degradation of agency authority and much else. Most of this anticipated malaise has never materialized. After a large number of NEPA lawsuits in the first few years, NEPA-related litigation declined sharply through subsequent decades. By the mid-1980s it was apparent delays in implementing NEPA-related agency programs were more often caused by the frequently tedious process of acquiring the stockpile of information thought necessary to bulletproof EISs and by the procedural intricacies required than by litigation. "The federal courts," writes judicial scholar Lettie McSpadden, "came to treat the writing of environmental impact statements as a paper exercise. They generally ruled in favor of government projects as long as the EIS requirement had been observed" (McSpadden, 1997, p. 175). By 1997, for instance, the CEQ estimated that 102 NEPA cases were initiated against federal agencies and only 2 resulted in injunctions against agency actions (Council on Environmental Quality, 1997a: p. 355). About half the lead plaintiffs in NEPA suits against federal agencies are public interest and citizen organizations, a large proportion of which represent environmental and conservation interests. But many public interest organizations have also been created to advance ideologically conservative resource agendas aligned with extractive industries. While few federal programs are ultimately cancelled by agency action or judicial injunction arising from the EIS process, considerable expense may be imposed on the participants by procedural delays and litigation, or by the need for affected private interests (such as extractive industries) to produce the necessary technical and scientific data required by the EIS process.

Moreover, agencies have discovered they can avoid much of the procedural delay and data harvesting involved in a formal EIS by substituting an Environmental Assessment. EAs are now the most common way in which agencies perform a NEPA analysis (CEQ, 1997b, p. 19). For instance, the U.S. Forest Service estimated in the mid-1990s that it issued 12,000 to 15,000 decisions for which NEPA documentation was required and most of these were documented as EAs (U.S. Senate, 1996).

The EAs and the closely related FONSI require less rigorous scientific analysis, the consideration of fewer policy alternatives, and little if any public involvement. Agencies resort to EAs and FONSIs when they make a determination that a proposed NEPA-related action will have relatively minor impacts to be mitigated by proposed measures or will have insignificant environmental consequences. These alternative strategies are criticized because they allegedly prevent desirable public access to agency decisions and because the cumulative impact of many minor agency decisions may be substantial environmental consequences appropriate for a formal EIS document. Moreover, agencies commonly provide little information about the extent to which promised mitigation measures are implemented and effective. All of which implies constricting public participation and availability of information involving NEPA-related agency decisions.

Discussions about diminishing public involvement in the NEPA process, however, would sound incongruous to governors and extractive industries. Many less partisan observers, moreover, claim the NEPA process (in the words of one survey) "encourages agencies and decision makers to acknowledge potential environmental consequences to the public, opening up the process (Canter & Clark, 1997, p. 316). What can reasonably be concluded from such dissonant information? First, there is a difference between public access to information and public influence upon the decisions to which that information is relevant. That public access to NEPA-related information has expanded enormously seems undeniable and the corollary—that such access mobilizes stakeholders and advocacy organizations about agency actions in ways otherwise less likely—seems equally plausible. But the most apparent influence exerted by the public—whether organized or not—upon agency behavior is mediated through judicially imposed rules about process or substance involved in collecting information for NEPA-related decisions. Second, there is considerable anecdotal and impressionistic information that agency policymakers consider public viewpoints, think about the environmental implications of their decisions, or are otherwise sensitive to publicly related issues through NEPA-related procedures. But convincing evidence that agency opinions have been significantly altered as a result of public awareness is more elusive. Third, complaints about negative impacts of public involvement in the NEPA process usually turn out to be complaints about agency decisions responsive to the federal judiciary, or about the time and cost created by agency NEPA procedures imposed in making policy decisions, or about agency determinations whose relationship the public aspects of the NEPA process are unclear. In short, it's a lot easier to attribute to NEPA a significantly increased public involvement in federal agency policymaking affecting the environment than to characterize its impact.

"A framework for collaboration?"

There is enough fragmentary evidence to suggest federal agencies frequently make different, and often ecologically better, decisions than they would otherwise have made as a result of the EIS process. It is inherently difficult to quantify the results, especially when the changes inspired by EISs in agency policies may be highly nuanced (incremental environmental improvement at the margin). The CEQ has struggled to characterize the impact and was content to conclude on NEPA's twenty-fifth anniversary that the NEPA's most enduring legacy has been "a framework for collaboration between federal agencies and those who will bear the environmental, social, and economic impacts of agency decisions"—a verdict susceptible to diverse interpretations (CEQ, 1997a, p. ix).

In one respect, at least, the CEQ's verdict seems reasonable. The EIS process has substantially increased the flow of information and institutional viewpoints across federal agencies affecting environmental quality and, on some occasions, promotes the development of formalized coordinating committees and other bureaucratic entities intended to facilitate the integration of this information in agency policymaking. One of the most significant structures of this kind results from the EPA's statutory responsibility to review and comment on all EISs produced by federal agencies. This review process has promoted the development of better environmental monitoring and reporting data in EPA and its diffusion to other federal agencies. At the same time, consultation and collaboration between agencies concerning EIS issues is not as predictable or successful as the CEQ suggests. Consider, for example, the dour verdict reached by the General Accounting Office about agreements intended to facilitate interagency consultation and review about EISs of common concern:

> Interagency agreements (1) have not been lived up to by agencies in the past, (2) are generally not enforceable by outside parties, (3) do not provide a basis for common approaches among all agencies. Also, federal land management and regulatory agencies sometimes do not work efficiently together to address issues that transcend their boundaries and jurisdictions. In addition, the environmental and socioeconomic data gathered by federal agencies are often not comparable. . . . (U.S. Congress, General Accounting Office, 1997, p. 10)

NEPA's exponents frequently assert that a significant enrichment of environmental information and communication between state and federal governments has resulted from the EIS process. Evidence of consultation about environmental policy across federal boundaries is abundant. Frameworks for collaboration, of course, are not necessarily manifestations of productive or successful collaboration. Many states still complain they are more often consulted, or informed, about their viewpoints on EIS related policies than conceded significant influence in the process.

The extent to which this framework of collaboration involves the public is also problematic. Almost all federal agencies conform to the CEQ's requirements that draft and final EISs be available for public review and comment. However, since EAs have largely replaced formal EISs, even the CEQ has complained about the resulting constriction of opportunity for public involvement in NEPA-related agency decision making. The extent and impact of public involvement in EIS preparation itself seems to be agency specific. The U.S. Geological Survey and the Corps of Engineers, for instance, are often praised for their public involvement strategies. The U.S. Forest Service, which usually produces more annual EISs than any other agency, is often criticized for its clay pigeon style of public involvement: public review is largely confined to consideration and comment on the preferred alternative(s) in a draft EIS plan—a style often discovered among other agencies as well (Welles, 1997).

Evaluating the EIS process

The criterion for assessing environmental policy instruments provided in chapter 1 provide an informative method to summarize the implications of the available data concerning the impact of the EIS process on federal governmental decision making. However, as most NEPA surveys ritually caution, inferences must be constrained in light of the statistically and conceptually limited data, the reliance on case studies, and the sometimes anecdotal quality of the available information.

Environmental effectiveness

Considering the impact of the EIS process upon environmental quality, several conclusions seem reasonable:

- The EIS process has significantly improved the quality of scientific information and the standards of professional practice utilized by federal agencies in evaluating the environmental impact of their decision making. In particular, this has greatly enlarged an understanding of the ecosystem consequences of federal policymaking, stimulated research on environmental indicators and increased environmental monitoring, and promoted investigation of largely unexplored issues such as the cumulative impact and socioeconomic consequences of federal environmentally-related policies.

- In some documented instances—such as the development of federal hydroelectric infrastructures, commercial nuclear power plant siting, water quality characterization, environmental monitoring and water resource projects—the EIS process has compelled major changes in environmentally important policies (see National Academy of Science, 1990: 231–234). In addition to these documented examples, it is widely believed that an agency's preparation of EIS documents often has a multiplier effect by influencing, often in ways difficult to document, subsequent decision making on related matters.

- Most federal agencies now routinely prepare technically competent EISs conforming to CEQ's regulations and judicially imposed guidelines, but the extent to which these EISs are carefully and routinely considered throughout the planning process remains speculative. Not surprisingly, the CEQ's Twenty-Fifth Anniversary review of NEPA noted that "the greatest hindrance to effective implementation of NEPA by agencies is when, and how, the NEPA process is triggered . . . Almost all the study participants support the need to apply NEPA earlier in the planning process . . . (Council on Environmental Quality, 1997a, p. ix). Among the impediments to EIS integration in agency planning processes, observers note that many, if not most, EIS statements are prepared by consultants, that agencies often do not clarify how they define a significant decision affecting the environment, and that agency resources allocated for the NEPA process are often limited and, in recent years, diminishing. Additionally, few agencies monitor the long-term consequences of actions they initiate to mitigate the environmental impact of a policy or project as promised in an EA or FONSI statement.

Economic efficiency

No indictment of NEPA is more familiar than the complaint that the EIS process is customarily too costly, time-consuming, and subversive to socially or economically desirable federal enterprises—in short, EIS benefits often cannot justify their cost.

- Little evidence suggests that EIS procedures impose excessive costs or delay upon agency decision making, or demonstrates that a significant number of federal projects are casualties of NEPA-imposed reviews. One recent survey of twelve federal agencies concluded: "In *none* of the . . . agencies reviewed. . . . did NEPA emerge as the principle cause of excessive delays or costs. Instead, the NEPA process was often viewed as a means by which a wide range of planning and review requirements were integrated (Smythe & Isber, 2002: Executive Summary). The Federal Highway Administration, a major EIS producer, estimates that EISs represented about 10 percent to total project costs and about 28 percent of total planning time. However, agencies find it difficult to assign costs and time specifically to the EIS process. While one survey of state transportation agencies revealed that the average time for the NEPA process had increased to 4.4 years, the process also included many elements not unique to NEPA (U.S. Department of Transportation, Federal Highway Administration, 2003). In general, while EIS preparation may slow the project planning (especially water resource projects) in some agencies, it is seldom considered the major reason for project delay.
- It is customarily difficult to estimate the benefits created by EIS procedures because many benefits involve the preservation of environmental amenities or the prevention of cumulative environmental damage not presently monetizable. Although benefit/cost analysis is sometimes included in agency EISs and EAs related to specific projects or policies, estimates of the benefit/cost value of the EIS process itself is almost never calculated.

Political and administrative efficacy

How well the EIS process works politically involves a consideration of whether the procedure produces politically viable solutions, facilitates the reconciliation of conflicting interests, and promotes support for the policies resulting for EIS assessment. Many NEPA evaluations, however, do not clearly distinguish between judgments about political and administrative efficacy—to many agency managers, for instance, whatever facilitates the implementation of projects or compliance with agency policies is politically efficacious for the agency.

- The EIS process has significantly enlarged the number of interests which have become informed and active in agency decision making affecting the environment. Many agencies still invest minimal effort in promoting public involvement in the NEPA process, but many others have aggressively encouraged public participation not only in EIS review but in the preparation of EAs where public participation is not a formal requirement. Many observers believe the significant decline in the frequency of NEPA litigation in recent decades is evidence that intensified public involvement activities have promoted greater acceptance of EIS-related policies. However, agencies are also frequently criticized for their lack of effort at public communication and education (BLM National Wilderness Campaign). Perhaps the most obvious impact of the public participation requirements in NEPA is to alert affected groups and individuals about the environmental implications of policies and, thus, to create a tripwire that mobilizes affected interests which otherwise might not be involved.

- Since most of the EISs and EAs are produced by five agencies, it is difficult to determine how broadly the EIS process has permeated the greater federal bureaucracy. NEPA procedures are practiced most often by the U.S. Forest Service, the Bureau of Land management, Department of Housing and Urban Development, the Corps of Engineers and the Federal Highway Administration.

- Most surveys of agency NEPA staff and others affected by the EIS process believe that EISs have encouraged greater acceptance of related agencies policies and have facilitated the subsequent implementation of those policies. At the same time, NEPA staff often complain about reduced budgets and personnel and the need for more training. Additionally, many NEPA offices believe greater resources need to be invested in the monitoring and collection of environmental indicators with which to make informed judgments about the success of mitigation activities involved in EAs and FONSIs.

- Congress has sometimes intervened in the EIS process to influence the manner in which specific projects or decisions are treated but generally has resisted major attempts to modify the generic procedures. Despite frequent Congressional criticism of the EIS process, Congressional interventions have been relatively rare. A more significant political intervention has been practiced by recent Presidential decisions to exclude public land use designations from EIS procedures.

Technological innovation

- Despite a need for better data, monitoring and evaluation related to projects affected by EIS preparation, one undisputed result of NEPA procedures has been a significant enrichment and expansion of the data base, scientific research, and professional training associated with environmental management. EIS preparation, for example, has undoubtedly been a major reason for the development of the relatively new science of ecosystem analysis and for increased research in the cumulative effects of environmental interventions.
- EIS preparation has stimulated research and practices that facilitate the integration of scientific data into agency planning processes. NEPA's mandate for EIS preparation has focused attention in many agencies upon procedures that effectively integrate the science and policy aspects of project planning, particularly at early stages in policy development. Frequently, the review of EIS documents has also illuminated the inadequacy of such integration in many agencies as well.

References

Andreen, W. L. (1990). In pursuit of NEPA's promise: The role of executive oversight in the implementation of environmental policy. *Land Use and Environmental Law Review, 21*, 559–616.

Andreen, W. L., & Blaug, E. L. (1993). Use of the environmental assessment by federal agencies in NEPA implementation. *Environmental Professional, 15*, 57–65.

Andrews, R. N. L. (1976). *Environmental policy and administrative change.* Lexington, MA: D. C. Heath.

Bartlett, H. (2000). Is NEPA's substantive review extinct, or merely hibernating? *Tulane Environmental Law Journal, 13,* 411–450.

Bear, D. (1992). The National Environmental Policy Act. *Intergovernmental Perspective,* Summer, 17–19.

Bear, D. (1993). NEPA—Substance or merely process?: Toward a stronger national policy on the environment. *Forum for Applied Research and Public Policy, 8.*

Bear, D. (1995). The National Environmental Policy Act: Its origins and evolution. *Natural Resources and the Environment, 3,* 3–6, 69–73.

Belsky, M. H. (1998). Environmental policy and NEPA: Past, present and future. *Ecosystem Health, 4,* 134–136.

Blaug, E. A. (1993). Use of the Environmental Assessment by Federal Agencies in NEPA Implementation. *Environmental Professional, 15,* 57–65.

Caldwell, L. K. (1982). *Science and the National Environmental Policy Act: Redirecting Policy Through Procedural Reform.* Tuscaloosa: University of Alabama Press.

Caldwell, L. K. (1997). Implementing NEPA: A non-technical political task. In R. Clark & L. Canter (Eds.), *Environmental Policy and NEPA: Past, Present and Future* (pp. 25–50). Boca Raton, FL: St. Lucie Press.

Caldwell, L. K. (1998). Beyond NEPA: Future significance of the National Environmental Policy Act. *Harvard Environmental Law Review, 22,* 205.

Calvert Cliffs Coordinating Committee, Inc. v. *United States Atomic Energy Commission* (1971), 449 F.2d 1109 (D.C. Cir), 146 U.S. App. D.C. 33.

Canter, L., & Clark, R. (1977). NEPA effectiveness—A survey of academics. *Environmental Impact Assessment Review, 17.*

Clark, R., & Canter, L. (1997). *Environmental policy and NEPA: Past, present, and future.* Boca Raton, FL: St. Lucie Press.

Cohen, S. F. (1980). The effectiveness of NEPA: Reactions to environmental impact review. *Vance Bibliographie.*

Dennis, N. B. (1997). Can NEPA prevent "ecological train wrecks"? In R. Clark & L. Canter (Eds.), *Environmental policy and NEPA: Past, present, and future* (pp. 139–162). Boca Raton, FL: St. Lucie Press.

EIS: Digests of Environmental Impact Statements. Bethesda, MD: Cambridge Scientific Abstracts.

Environmental Law Institute. (1995). *Rediscovering the National Environmental Policy Act: Back to the future.* Washington, DC: Author.

Flint Ridge Development Co. v. Scenic Rivers Association of Oklahoma. 426 U.S. 776 (1976).

Friends of Fiery Gizzard v. Farmers Home Administration. 61 F. 3rd 501 (6th Cir 1995).

Fogleman, V. M. (1990). *Guide to the National Environmental Policy Act: Interpretations, applications, and compliance.* New York: Quorum Books.

Frye, K., Yates, E., & Batts, D. (2001, June). *A comparison of how different federal agencies comply with the National Environmental Policy Act.* Paper presented at the annual conference of the National Association of Environmental Professionals.

Hanley v. Kleindienst, 471 F. 2nd 823 (2nd Cir 1972), *cert. denied,* 412 U.S. 908 (1973).

Hays, S. P. (1987). *Beauty, health and permanence: Environmental politics in the United States, 1955–1985.* New York: Cambridge University Press.

Hill, W. W., & Ortolano, L. (1978). NEPA's effect on the consideration of alternatives: A crucial test. *Natural Resources Journal, 18,* 285–311.

Jesse, L. (1998). The National Environmental Policy Act Net (NEPAnet) and DOE's NEPA Web: What they bring to environmental impact assessment. *Environmental Impact Assessment Review, 18,* 73–82.

Lazar, R. (1978). The National Environmental Policy Act and its implementation: A selected, annotated bibliography. *Wisconsin Seminars on Resource and Environmental Systems.* Madison, WI: Institute for Environmental Studies.

Lindstrom, M. J. (2000). Procedures without purpose: The withering away of the National Environmental Policy Act's substantive law. *Journal of Land, Resources and Environmental Law, 20,* 245–268.

Lindstrom, M. J., & Smith, Z. A. (2001). *The National Environmental Policy Act: Judicial misconstruction, legislative indifference and executive neglect.* College Station, TX: Texas A&M Press.

Lynn, W. R. (1993). *Changing Patterns of water resource decision making in sustaining our water resources.* Washington, DC: National Academies Press.

Mandelker, D. (1984). *NEPA Law and Litigation.* New York: Clark Boardman Callaghan.

Manheim, B. S., Jr. (1994). NEPA's overseas application *Environment, 36,* 44.

McSpadden, L. (1997). Environmental policy in the courts. In N. J. Vig, & M. E. Kraft (Eds.), *Environmental policy in the 1990s* (3rd ed.). Washington, DC: CQ Press.

National Academy of Science. (1990). *Colorado River Ecology and Dam Management: Proceedings of a Symposium.* May 24–25, 1990, Santa Fe, New Mexico.

National Academy of Science. (1995). *Flood risk management and the American river basin: An evaluation.* Washington, DC: National Academies Press.

National BLM Wilderness Campaign. (2003). *Public Involvement at the BLM: A NEPA Report Card.* Retrieved from www.blmwilderness.org/nepa.mhml.

National Research Council, Commission on Geosciences, Environment, and Resources. (2000). *Downstream: Adaptive management of Glen Canyon dam and the Colorado River ecosystem.* Washington, DC: National Academy Press.

Oak Ridge National Laboratories. (2002). *ORNL Review, 25.* Retrieved from www.ornl.gov/ORNLReview/rev25–34/chapter6sb7.htm.

Peterson, R. C. (1971). An analysis of Title I of the National Environmental Policy Act of 1969. *Environmental Law Reporter, 1,* 50035–50055.

Plater, Z. J. B., Adams, R., & Goldfarb, W. (1992). *Environmental law and policy: Nature, law, and society.* St. Paul, MI: West Publishing.

Smythe, R., & Isber, C. (2003). *NEPA in the agencies, 2002: A report to the Natural Resources Council of America.* Retrieved from www.nrca.org.

Swartz, L. L. (1995). *Major cases interpreting the National Environmental Policy Act.* Retrieved from Battelle Memorial Institute, National Association of Environmental Professionals www.neap.org/NEPAWG/majorcas.htm.

Tzoumous, K., & Finegold, L. (2000). Looking at the quality of Draft Environmental Impact Statements over time in the United States: Have ratings improved? *Environmental Impact Statement Review, 20,* 557–578.

Taylor, S. (1984). *Making bureaucracies think: The environmental impact strategy of administrative reform.* Palo Alto: Stanford University Press.

U.S. Congress, General Accounting Office. (1997). *Forest Service Decision-Making: A Framework for Improving Performance.* Report No. GAO/RCED-97-71.

U.S. Congress, Senate. (1969). *National Environmental Policy Act of 1969.* 91st Congress, 1st Session. Report 91-296, pt. 14, at 19.

U.S. Congress, Senate, Committee on Environment and Public Works, Subcommittee on Superfund, Ocean, and Water Protection. (1991). *Amending the National Environmental Policy Act: Hearings.* 102nd Congress, 1st Session, July 11.

U.S. Congress, Senate, Committee on Environment and Public Works, Subcommittee on Environmental Protection. (1992). *National Wildlife Refuge System Management and Policy Act: Hearings.* 102nd Congress, 2nd Session, June 19.

U.S. Congress, Senate, Committee on Energy and Natural Resources, Subcommittee on Oversight and Investigations. (1995). *Application of the National Environmental Policy Act.* 104th Congress, 1st Session, June 7.

U.S. Congress, Senate, Committee on Energy and Natural Resources, Subcommittee on Oversight and Investigations. (1996). *Strengthening The National Environmental Policy Act: Hearings.* 104th Congress, 2nd Session, September 26.

U.S. Congress, House of Representatives, Committee on Resources. (1998). *Problems and Issues with the National Environmental Policy Act of 1969: Hearings.* 105th Congress, 2nd Session, March 18.

U.S. Department of Transportation, Federal Highway Administration. (2003). *Evaluating the performance of environmental streamlining: development of a NEPA baseline for measuring continuous performance.* Retrieved from www.fhwa.dot.gov/environment/strmlng/baseline/index.htm.

U.S. Environmental Protection Agency, Office of Federal Activities. EIS Page Lengths.

U.S. Executive Office of the President, Council on Environmental Quality. (1997a). *The National Environmental Policy Act: A study of its effectiveness after twenty-five years.* Washington, DC: Council on Environmental Quality.

U.S. Executive Office of the President, Council on Environmental Quality. (1997b). *Considering cumulative effects under the National Environmental Policy Act.* Washington, DC: Council on Environmental Quality.

U.S. Executive Office of the President, Council on Environmental Quality. (1998). *Environmental Quality: The 1997 Report of the Council on Environmental Quality, Part I.* Washington, DC: Council on Environmental Quality.

U.S. Executive Office of the President, Council on Environmental Quality. (1998). *Regulations.* Sec. 1508.9.

U.S. Executive Office of the President, Council on Environmental Quality. (1998). *Regulations.* Sec. 1508.13.

U.S. Executive Office of the President, Council on Environmental Quality. (1998). *Regulations.* Sec. 1508.3.

U.S. Executive Office of the President, Council on Environmental Quality. (1998). *Regulations.* Sec 1501.3.

U.S. Executive Office of the President, Council on Environmental Quality. (1998). *Regulations.* Sec. 1501.4.

Welles, H. (1997). The CEQ effectiveness study: Learning from our past and shaping our future. In R. Clark & L. Canter (Eds.), *Environmental policy and NEPA: Past, present, and future.* Boca Raton, FL: St. Lucie Press.

Zellmer, S. B. (1997). Sacrificing legislative integrity at the alter of appropriations riders: A constitutional crisis. *Harvard Environmental Law Review, 21,* 457–535.

CHAPTER TEN

Institutional and Technological Constraints on Environmental Instrument Choice

A Case Study of the U.S. Clean Air Act

DANIEL H. COLE AND PETER Z. GROSSMAN

Introduction

It is an article of faith among economists, legal scholars, and policymakers that economic forms of regulation such as effluent taxes and emissions trading are inevitably more efficient than traditional command-and-control regimes for environmental protection. Some suggest that command-and-control regimes are not only less efficient but inherently inefficient, implying that they naturally produce more social costs than benefits (see Tietenberg, 1985; Stewart, 1996).[1]

This chapter takes issue with the general portrayal of command-and-control environmental regulations in the economic and legal literature. The prevailing view—that command-and-control is inevitably inefficient or less efficient than alternative economic instruments[2] such as effluent taxes and marketable pollution permits—is inaccurate both as a matter of economic theory and experience. This chapter argues that command-and-control environmental regulations can be (and have been) nominally efficient (producing social benefits in excess of their costs); indeed, they can even be (and have been) more efficient than alternative economic approaches to regulation.

Standard economic accounts of command-and-control environmental regulations focus almost exclusively on compliance costs, ignoring technological and institutional constraints that can significantly affect the comparative efficiency of alternative environmental protection policies. When technological and institutional constraints are introduced into the analysis, command-and-control regulations do not appear to be invariably inefficient or necessarily less efficient than economic mechanisms for environmental protection. Indeed, in some cases, given (a) the marginal costs of pollution control, (b) technological constraints, and (c) existing institutions, command-and-control can be the most efficient means of achieving society's environmental protection goals.

The history of air pollution control under the U.S. Clean Air Act (CAA) demonstrates the vital importance of institutional and technological factors in environmental instrument choice. Viewed as an evolutionary process, occurring within an institutional and technological framework, it was nominally and probably relatively efficient for Congress to rely, in early years of federal air pollution control, on command-and-control regulations and, then, in more recent years to begin experimenting with efficiency-enhancing market-based alternatives.

The efficient evolution of federal air pollution control, 1970–1990

Many economists and legal scholars deplore the CAA's heavy reliance on what they consider to be blunt and often wildly inefficient and irrational regulatory instruments, such as national ambient air quality standards and technology-based emissions limitations (Orts, 1995, p. 1236, quoting Krier, 1974, pp. 323–25). But those instruments do not appear so irrational when one considers the institutional and technological constraints under which Congress was operating in 1970, when it established the Act.

The 1970 Clean Air Act: Instituting command-and-control

Congress possessed precious little information about the economic costs of pollution and the economic benefits of pollution control; existing estimates were very rough and subject to great uncertainties. What is worse, no one really knew how bad the country's air pollution problem was; ambient concentration levels, the rate of pollution emissions, and the level of necessary emission reductions to ensure minimally safe ambient concentration levels (however safe was defined) were hardly known. Existing estimates did not bode well. In 1940, for example, national sulfur dioxide emissions were estimated to be just under twenty million short tons. By 1970, estimated annual sulfur dioxide emissions had increased by 46 percent to just over thirty-one million short tons (EPA, 1997, pp. 3–13; Tables 3–4). It needs to be stressed, however, these estimates were rough at best, given the primitive state of emissions monitoring. It is quite likely that the existing estimates substantially understated actual emissions.

This general lack of air pollution information reflected existing technological constraints. As the National Air Pollution Control Administration (NAPCA) reported in 1970, existing air pollution monitoring equipment and analytical techniques were "not adequate to meet current and anticipated future needs in air monitoring, source testing, measurement of meteorological parameters, and laboratory research" (II Legislative History, 1974, pp. 1306–1307). Equipment for monitoring ambient concentration levels of various pollutants was lacking. Nationwide in 1970 there were 245 particulate matter (dust) monitors, 86 sulfur dioxide monitors, 82 carbon monoxide monitors, 43 nitrogen oxide monitors, and just one monitor for ozone (U.S. Environmental Protection Agency, 1997, D.3, Table D.1). These monitors and other available analytical equipment "lack[ed] accuracy, sufficient sensitivity to reflect progress in controlling air pollution, or the specificity needed to satisfy air quality criteria requirements."[3]

Individual point-source emissions monitoring was in an even more primitive state than ambient concentration monitoring in 1970. For the most part, the government had only one option: to rely on industry self-monitoring (to the extent possible) and reporting on emissions rates. Predictably, some industries were simply unwilling to provide the government with information on their emissions.[4] Others said (perhaps truthfully) that the information simply was not available (II Legislative History, 1974, p. 1312). Many sources did comply with government requests for information on their emissions; still, the government had no means of verifying the data it received.

Given the sense that air pollution was increasing, but the dearth of information about the extent of the nation's air pollution problems and the emissions reductions necessary to attain healthful air quality, Congress in the 1970 CAA rationally focused its attention on rapidly improving air quality, rather than questions of how it should be done or how much (see generally II Legislative History, 1974). Indeed, in more than 1,500 pages of legislative history that accompanied the Act, there was virtually no discussion of alternative approaches to regulation [beyond a single reference to effluent taxes during a hearing on May 27, 1970 (II Legislative History, 1974, pp. 1223–1224)]; Congress never contemplated effluent taxes or tradable permitting, though economists and environmental policy analysts were already advocating their use (see, e.g., Dales, 1968). This suggests (on a Public Choice account) Congress and the interest groups pressing for federal air pollution control legislation had little interest in efficiency-enhancing economic instruments. But it certainly reflects the real information constraints (particularly relating to emissions levels) that Congress faced in 1970.

Beyond information constraints, Congress was deeply concerned with the inadequate staffing of state and federal agencies charged with implementing air pollution regulations. According to a 1970 report to Congress by the Department of Health, Education, and Welfare (the federal agency with primary responsibility for environmental protection before the creation of the Environmental Protection Agency in 1970), control agencies were "in general . . . inadequately staffed. Fifty percent of state agencies [had] fewer than ten positions budgeted, and 50 percent of local agencies [had] fewer than seven positions budgeted." The HEW report estimated that state and local agency staffing would have to increase by 300 percent "to implement the Clean Air Act properly" (I Legislative History, 1974, p. 254). Meanwhile, according to Senator Muskie (chief sponsor of the CAA in the Senate), federal agency staffing needed to be tripled by 1973 to fully implement the Act (I Legislative History, 1974, p. 230). To fill this need, Congress in the CAA created a sizeable federal grant program to universities for training a new cadre of environmental protection experts (II Legislative History, 1974, p. 1296).

Because of these various institutional and technological constraints, it made sense for Congress in 1970 to take a public-property/regulatory approach to air pollution control. Specifically, by focusing on technology-based design standards for industrial sources and automobiles, the federal government could minimize its monitoring and staffing deficiencies. As noted earlier, so long as pollution control technologies were installed and operating, the government could be assured of some emissions reductions, even if it could not precisely measure them.

Several contributors to a volume published a year before the 1970 CAA was enacted recognized this advantage of command-and-control. George Hagevik wrote, "[t]he advantage of [direct regulation] is that it permits the government to take interim steps even though it has almost no idea of relevant measurements" (Hagevik, 1969, p. 178). The economist Harold Wolozin noted that "[f]ormidable detection and monitoring problems are implicit in effluent fee schemes, a problem compounded by the primitive state of technology in these areas" (Wolozin, 1969, pp. 31, 39). Wolozin quoted from a presentation by William Vickrey at the 1967 annual meeting of the Air Pollution Control Association: "The real problem which advocates of effluent charges must face is the problem of metering, or of estimating in some way the amount of effluent actually generated by various emitters. Here the problem of air pollution is seen to be a particularly difficult one in that the number of small emitters and of the emitters difficult to meter effectively is large and their contribution to the problem is too great to be ignored" (Wolozin, 1969, p. 40). Paul Gerhardt agreed with Vickrey, Wolozin, and Hagevik that "a fee system could be exceedingly difficult and costly to administer. . . . [E]mission measurement technology is presently inadequate to meet the requirement that a regulatory agency be able to determine with some precision just how much an individual polluter is contributing to the atmospheric burden" (Gerhardt, 1969, pp. 162, 169).

Existing technological constraints rendered impracticable regulatory approaches such as effluent taxes and tradable permits that depend on regular and precise point-source monitoring of emissions. They may have been efficient in theory—perhaps more efficient than the command-and-control mechanisms that Congress codified in the 1970 CAA—but only if monitoring costs were ignored. Unfortunately, many economists and legal scholars who write about instrument choice in environmental protection do just that: ignore monitoring costs. As Clifford Russell, Winston Harrington, and William J. Vaughn have noted, most environmental policy analysts tend to assume "perfect (and incidentally, costless) monitoring" (Russell, 1986, p. 3). This may be convenient analytically, but it is obviously unrealistic and skews perceptions of the relative efficiency of effluent taxes and tradable permits versus technology-based command-and-control regulations. Given that regular and precise point-source monitoring is necessary to accurately assess taxes or determine compliance with emissions quotas (pursuant to a tradable permit scheme), and given the infeasibility of monitoring emissions from tens of thousands of individual smokestacks and millions of automobile tailpipes in 1970, it was rational—indeed, efficient—for Congress to rely on technology-based command-and-control standards.

The 1977 Clean Air Act Amendments: Retaining the regulatory status quo

Institutional and technological circumstances changed significantly, though not decisively, between 1970 and the first set of major amendments to the Clean Air Act in 1977. On the institutional front, the Environmental Protection Agency, which was created in 1970 to implement the CAA, showed a surprisingly rapid learning curve for a fledgling agency that had virtually no economic expertise at the outset. Indeed, before it was a year old, the agency's first administrator, Russell Train, testified before Congress in favor of President Nixon's proposals for effluent taxes on sulfur dioxide emissions and lead additives in gasoline (see III Legislative History, 1978, p. 2541). Congress did not enact either proposal, but is it interesting to note that President Nixon did not propose the tax on sulfur dioxide emissions as a preferable alternative to technology-based standards; rather, he proposed it as a second-best option because no technologies were then available for controlling sulfur dioxide emissions.[5]

By the mid-1970s, the EPA was implementing its own policies designed to introduce some much needed flexibility into the regulatory system, and thereby reduce compliance costs for regulated industries (see Liroff, 1986; Hahn & Hester, 1989, p. 361). These innovations demonstrated the agency's increasing competence on economic issues in environmental regulation and reflected, in part, a substantial alleviation of agency staffing deficiencies. By 1979, the federal EPA had more than 500 employees nationwide working exclusively on clean-air programs. State and local governments, meanwhile, devoted more than 6,500 personnel to air pollution control (National Commission on Air Quality, 1981, p. 94, Table 4). Whether this level of staffing was sufficient to meet the increased monitoring, data-collection, and record-keeping needs of effluent tax of tradable permit regimes is, however, unclear.

Technological constraints on effluent taxes and tradable permits were also marginally reduced by the time Congress amended the CAA in 1977. By then, the total number of ambient concentration monitors for criteria pollutants had increased by more than a factor of six;[6] the number of particulate matter monitors rose from 245 (in 1970) to 1,120 (in 1975); ozone monitors increased from 1 to 321; and so on (U.S. Environmental Protection Agency, 1997, D.3, Table D.1). Moreover, the quality and reliability of the monitoring equipment and analytical techniques for data interpretation had improved, though evidently not enough to make tradable permitting or effluent taxes feasible alternatives to command-and-control. A 1977 report by the National Research Council "identified the lack of statistical rigor in the design and analysis of most environmental monitoring networks" (CEQ, 1990, p. 56, citing National Academy of Sciences, 1977). The first continuous emission monitoring system (CEMS) became available in 1975 (see Janke, 1993, p. 2). This marked an important step toward making effluent tax schemes and tradable permit programs feasible. But in 1977 monitoring technologies still were not sufficient to permit the cost-effective replacement of command-and-control regulations with economic instruments. As Marc Roberts wrote in 1982:

> "[w]hen economists discuss such matters [as emissions trading] they sometimes talk as if monitoring devices were widely avail-

able to cheaply and reliably record the amount of all pollution emissions. If that were the case decisions about whether a source had curtailed its pollution by the promised amount and whether a new source was emitting no more than the tradeoff transaction implied could be left to straightforward data gathering by an enforcement agent. Unfortunately, such monitoring devices typically are not available." (Roberts, 1982)

Given the continuing infeasibility of tradable permitting and effluent taxation, stemming primarily from technology constraints, Congress in the 1997 CAA Amendments rationally left the existing command-and-control system in place. And the two major new programs Congress added to the Act in 1977—the nonattainment area rules and the Prevention of Significant Deterioration program—were based on the same command-and-control model. The 3,400 pages of legislative history accompanying the amendments reveals Congress was only marginally more interested in effluent taxes and tradable permits in 1977 than it had been in 1970, even though they had a great deal more information about pollution levels, necessary reductions, and regulatory alternatives; at least the information was available to any legislator who chose to be informed.[7]

The continuation of command-and-control in the 1977 CAA Amendments is usually explained from a Public Choice perspective: neither Congress, the EPA, the regulated industries, nor environmental groups favored a switch to alternative and less costly regulatory approaches (see, e.g., Keohane, Revesz, & Stavins, 1997). But our analysis suggests Congress's decision to stick with command-and-control in 1977 was rational given the continuing infeasibility of effluent taxes and tradable permits, due to continuing technological constraints. There was also a sense in Congress that the existing program was not overly expensive. National expenditures for air pollution control in 1975 were $15.7 billion, amounting to only about one percent of the economy's total output of goods and services. Meanwhile, air pollution was conservatively estimated to cost the economy (in health and material damage) more than $25 billion annually (IV Legislative History, 1978, pp. 3050–1).

By 1977, Congress had some idea that its command-and-control regulations were improving air quality. Based on the information provided by increased emissions monitoring, the government was able to establish a clear downward trend in criteria pollutant emissions (except nitrogen oxide). Sulfur dioxide emissions, for example, declined by an estimated 17 percent, from just over thirty-one million short tons in 1970 to just under twenty-six million short tons in 1980 (EPA, 1997, pp. 3–13, Table 3–4).

What Congress did not know in 1977 was that the existing command-and-control system was actually producing sizeable net social benefits. But we know that now. A 1982 study determined the Clean Air Act produced net social benefits of more than $26.3 billion between 1970 and 1981 (Freeman, 1982).[8] A more recent study by the EPA estimates the CAA produced (inflation-adjusted) net benefits of $909 billion between 1970 and 1980 (U.S. EPA, 1997, p. 56, Table 18). This cannot be over-emphasized: the presumptively costly command-and-control system of the Clean Air Act was (and is still) producing huge net social benefits. Again, Congress did not know this when it enacted the 1977 Amendments, but there was a sense that the Act was not imposing excessive costs on society

So, what impetus was there for Congress to make a costly switch to new and as yet untested regulatory approaches? No studies available to Congress in 1977 (and no studies completed since then) support the proposition that replacing the existing command-and-control system with alternative economic instruments would have produced larger net social benefits in 1977, after accounting for the costs of transitioning to the new system, including increased monitoring costs.

The 1990 CAA Amendments: An incremental shift toward economic instruments

In the 1990 CAA Amendments, Congress took environmental policy in a new direction. A new subtitle of the Act, designed to deal with the problem of acid rain, departed from the traditional command-and-control framework by creating a market for tradable pollution permits. This was the largest-scale experiment yet undertaken with tradable permits. Senator Baucus, a primary sponsor of the 1990 CAA Amendments, clearly expressed the experimental nature of the program: "Many of the provisions in this bill are new ideas. Allowance trading in the acid rain title. We think it will work. We thought it through as well as we could. We are not sure it will work as well as we had intended. Therefore, there will be many adjustments, modifications and refinements as we work with an experiment with the acid rain portion of this bill" (Legislative History, 1993, Vol. 1, p. 1142). Other legislators and President Bush referred to the emissions trading scheme as innovative, novel, and never tried . . . before (Legislative History, 1993, Vol. 1727, 1144, 1269; Vol. 2, pp. 2550, 3389; Vol. 4, pp. 4857, 6391).[9]

And the experiment has worked so far better than anyone expected. By November 1995, 23 million SO_2 allowances worth $2 billion had been transferred in more than 600 market transactions.[10] The first allowances sold in 1992 for between $250 and $400 a piece; the average fell to $68 in 1996, but rebounded to $107 in 1997 (Percival, 1996, pp. 831–32; Air Pollution, 1997). In addition to these transfers, many sources have saved and banked excess allowances for future use when further reductions are required beginning in the year 2000. The result has been a greater than expected reduction in SO_2 emissions and a 10 to 25 percent reduction in acid precipitation in the Northeast (Swift, 1997, p. 17). Total emissions in 1995 were five point three million tons, 39 percent below the legislatively set ceiling and more than 50 percent below 1980 emission levels (Zacaroli, 1996). By 1996, national SO_2 emissions had fallen below estimated 1940 levels. Moreover, the introduction of the emissions trading program in the 1990 CAA Amendments has led to a significant rise in the rate of emissions reductions. Under command-and-control, between 1970 and 1990, SO_2 emissions declined at an average annualized rate of 1.3 percent. In the first six years after Congress enacted the acid rain program, sulfur dioxide emissions declined at an annualized rate of 2.9 percent (calculations based on figures in EPA, 1997, pp. 3–13, Table 3–4). In other words, the cap-and-trade approach of the 1990 acid rain program was reducing sulfur dioxide emissions more effectively than the preceding system of command-and-control. And it was doing so more efficiently.

Economically, the acid rain program's SO_2 trading program has been "a terrific bargain" (Percival, 1969, p. 832), producing substantial net benefits for society (Burtraw, 1997, p. 26). The lowest estimates of its annual health benefits—twelve billion—are four times higher than the highest estimates of annual program costs (Percival, 1969, p. 832).[11] A key question, of course, is whether the acid rain program's emission trading mechanism increased net benefits over what they would have been under direct regulation. Total cost savings are difficult to estimate, but must be substantial. Consider that just four utilities (Central Illinois Public Service, Illinois Power Company, Duke Power, and Wisconsin Electric Power Company) have estimated their aggregate savings from purchasing allowances rather than installing scrubbers at $706 million (U.S. General Accounting Office, 1994, pp. 33–34). This figure is not far below the total annual costs of compliance with Phase I SO_2 emission-reduction requirements, estimated at $836 million (Energy Information Administration, 1997, p. 12).

Although Congress's adoption of an emissions trading program in the acid rain provisions of the 1990 Clean Air Act Amendments represented a new direction in environmental policy, it was not simply a matter of Congress finally realizing what economists had known for decades: that tradable permits are more efficient and effective than technology-based standards. The move to tradable permits was facilitated by institutional, technological, and economic developments between 1977 and 1990—specifically, the rising marginal costs of air pollution control, the perception that those costs were overtaking social benefits of the Act, the EPA's increasing economic expertise, and most importantly the development of reliable and cost-effective CEMS.

Between 1977 and 1990 emissions of all criteria pollutants fell. The average decline in emissions was 24 percent (but only 11.2 percent if lead emissions are excluded).[12] As a consequence, ambient air quality improved significantly. National ambient concentrations of criteria pollutants fell between 1981 and 1988 by an average of 22.6 percent (only 10.6 percent if ambient lead concentrations are excluded).[13] These substantial improvements in air quality came at a higher cost to society, however, as the total annualized costs of air pollution control increased (in constant 1990 dollars) by 63.5 percent, from $15.9 billion in 1977 to $26 billion in 1990 (U.S. Environmental Protection Agency, 1997, A-16, Table A.9).[14]

This did not mean the CAA was producing net social loses; to the contrary, it was still yielding substantial (indeed, growing) net social benefits. Between 1970 and 1975, the Act produced estimated net, inflation-adjusted benefits—defined as the "mean monetized benefits less annualized costs for each year"—of $341 billion (U.S. Environmental Protection Agency, 1997, pp. 55, 56, Table 18). From 1975 to 1980, estimated net benefits nearly tripled to $909 billion; they rose to $1.13 trillion by 1985, and stood at $1.22 trillion in 1990 (U.S. Environmental Protection Agency, 1997, pp. 55, 56, Table 18).[15] But Congress did not know all this when it enacted the 1990 CAA Amendments.[16] Indeed, there was a fairly widespread perception that the CAA was becoming too costly. The ten thousand pages of legislative history that accompanied the 1990 Amendments are replete with expressions of concern and debates over the respective costs and benefits of new and existing air pollution control programs (see I Legislative History, 1993, pp. 731, 1058, 1091, 1179, 1193; VI Legislative History, 1993, pp. 4829, 7187, 9736–9755, 9826, 9867). And because perception is at least as important as fact in the legislative process, congressional interest in efficiency-enhancing policies may have increased as legislators and others perceived that additional increments of air pollution control could only be obtained under the existing program at a net cost to society.

But there is more to the story behind the 1990 CAA Amendments than congressional perceptions of costs and benefits. Important institutional developments since 1977 provide reason to believe federal and state agencies could successfully implement and enforce tradable permit and other alternative regulatory programs. The federal EPA certainly had grown more open to the use of alternative regulatory mechanisms, such as tradable permits, by 1990. The agency had been preparing economic analyses, Regulatory Impact Analyses (RIAs) of its major regulations since its inception.[17] And it had grown comfortable with cost-benefit analysis as a tool of environmental policy. As a 1987 EPA report noted, "[e]nvironmentalists often fear that economic analysis will lead to less strict environmental regulations in an effort to save costs, but our study reveals that the opposite is just as often the case" (U.S. Environmental Protection Agency, 1987, p. 2). In order to meet its obligations to perform economic analyses, the agency by 1990 was employing more than one hundred economists, which gave it greater economic expertise than any other federal health and safety agency (Davies & Mazurek, 1998, p. 32).

The EPA's economic expertise and comfort with economic instruments also increased thanks to a small-scale and temporary but highly successful experiment in tradable rights to lead content in gasoline. As part of its phase out of leaded gasoline, from 1982 to 1987 the EPA allowed trading in rights to add lead to gasoline. Refiners producing gasoline with a lower lead content than mandated by federal standards could sell or bank the excess lead content. Smaller refiners that could not afford to meet the federal lead content standards were able to reduce their compliance costs by purchasing lead rights instead of reducing lead content. The program proved reasonably successful. "[T]he market in lead rights was very active, and this activity generally increased throughout the life of the program." According to some sources, "[t]he cost savings from lead rights trading and banking . . . amounted to hundreds of millions of dollars" (Hahn & Hester, 1989, pp. 361, 380–391). What made lead-content trading technologically feasible when emissions trading was not, was the fact that lead-content—unlike fugitive emissions—remains measurable at any time in the gasoline, facilitating compliance enforcement. Nevertheless, this successful experiment with trading in lead-content in gasoline undoubtedly increased the EPA's confidence with tradable permitting as a useful policy tool.

Just as the EPA was becoming more familiar with and expert in dealing with economic issues in environmental protection, state environmental agencies were becoming better staffed and equipped. Between 1974 and 1992, state air quality expenditures more than doubled from $249 million to $516 million (in constant 1992 dollars) (Davies & Mazurek, 1998, p. 32). This may have alleviated federal concern about the states' abilities to monitor and enforce federal air pollution control programs.

Even more important for the 1990 Amendments than institutional developments, were technological changes that more than any other factors made tradable permitting (and effluent taxes) administratively feasible. Continuous emission monitoring systems, the first of which appeared two years before Congress enacted the 1977 CAA Amendments, had become widely available and affordable by 1990. As James A. Janke wrote in 1993, "CEM systems . . . advanced considerably over the past fifteen years, with improved sampling techniques, analyzers, and data processing systems being integrated to meet the challenges posed by new requirements" (Janke, 1993, p. 8). By 1991 the U.S. government was requiring continuous monitoring at twenty-four categories of sources subject to New Source Performance Standards under the CAA. In addition, CEMS were required for all electric power generators regulated under the CAA's acid rain program (Janke, 1993, pp. 12–13, Table 2–1). CEMS were not yet available for all air pollutants or for all categories of sources,[18] but they made tradable permitting a feasible (i.e., cost-effective) and in some cases preferable (i.e., more efficient) policy choice for certain combinations of pollutants and sources.

With increased information about market mechanisms for environmental protection, the innovation of new technologies (particularly CEMS) to adequately monitor point-source emissions, improved funding of state agencies, the development of agency economic expertise, and rising concern about the costs of air pollution control, it made sense that Congress, in the 1990 CAA Amendments, would begin experimenting on a larger scale with alternative regulatory approaches, such as emissions trading. Whether or not the CAA was still producing net social benefits (it was, as we shall see), the reasons for preferring command-and-control—particularly concerns over monitoring and enforcement—were waning. At least with respect to certain air pollution problems, the theoretical efficiency advantages of tradable permits and effluent taxes could finally be realized.

One frequently overlooked point about the acid rain program is Congress's insistence, in the text of the statute itself, on continuous monitoring of sulfur dioxide emissions, "to preserve the orderly functioning of the allowance system, and . . . ensure the emissions reductions contemplated by this [program]" (42 USC & 4651k(a), 1997). This statutory language reflects two critical perceptions. First, "[u]nlike other control requirements of the CAA, utility emissions of SO_2 and NO_x are capable of verification in a cost-effective manner through the use of continuous emission monitors." Second, "[t]he requirements for CEMS is the linchpin in this title for without good emissions data, a problem that has hampered enforcement of the Act to date, no allowance or emissions trading scheme can affectively [sic] operate" (I Legislative History, 1993, p. 1040). The clear implication is that absent the technical capability to precisely monitor emissions in a cost-effective manner, tradable permitting cannot be said to be feasible, let alone preferable to direct regulation. It is the existence of cost-effective technologies for measuring SO_2 emissions that has made the emissions trading program workable.

Despite the success of the acid rain program's emission trading program, it is important to bear in mind (as Congress apparently did) that reliable and cost-effective CEMS are not available for all combinations of pollutants and sources regulated under the CAA. This may explain why Congress did not make wholesale changes in the 1990 Amendments, but restricted its use of emissions trading to only one program. As the New Institutional Economics might predict (see, e.g., North, 1990), Congress's policy shift away from the pure public-property/regulatory approach of command-and-control towards the mixed-property/regulatory approach of tradable permitting has been deliberate, incremental, and inconsistent.

Implications and conclusions

According to the standard economic account, command-and-control regulations are inefficient, and should grow increasingly inefficient over time as additional pollution control becomes more costly. But the best available numbers on the Clean Air Act do not bear this out. Something must be wrong either with the numbers, the standard argument, or both.

The numbers undoubtedly are imperfect. Estimations of environmental costs and, especially, environmental benefits are fraught with uncertainty and subjectivity (especially in the valuations of non-priced goods and bads and the selection of discount rates). But it would take a clever accountant indeed to make the numbers show that the Clean Air Act imposes net costs on society. As the EPA has explained, "the benefits of the Clean Air Act and associated control programs [between 1970 and 1990] substantially exceeded costs. Even considering the large number of important uncertainties permeating each step of the analysis, it is extremely unlikely that the converse could be true" (U.S. Environmental Protection Agency, 1997, ES-9). If EPA's mean estimate of net benefits, 1970–1990 (compounded at five percent)—$21.7 trillion (U.S. Environmental Protection Agency, 1997, 56, Table 18)—were too high by a factor of 100, the CAA still would have yielded $217 billion dollars in net social benefits.[19] So, even if there is something terribly wrong with the numbers, there must also be something wrong with the claim that command-and-control is an inherently inefficient policy tool. Specifically, standard economic accounts of the comparative efficiency of alternative regulatory regimes are insensitive to institutional and technological contexts. Most importantly, they tend to assume "perfect (and incidentally, costless) monitoring" (Russell, 1986, p. 3) or they assume that monitoring costs are the same regardless of the control regime that is chosen. As we have shown in this chapter, both of these assumptions are unrealistic, and they skew comparative cost-benefit analyses of alternative regulatory instruments. When institutional and technological costs are considered, command-and-control regulations appear neither inherently inefficient nor invariably less efficient than theoretical economic approaches, such as effluent taxes or emissions trading schemes. Indeed, in some cases—such as those involving very high monitoring costs—command-and-control can be more efficient than market mechanisms.[20]

However, our goal in this study has not been to question the efficiency enhancing potential of effluent taxes or emissions trading programs. Indeed, as we have seen, the 1990 CAA Amendment's large-scale experiment with emissions trading in the acid rain program has been an unmitigated success, producing greater than expected emissions reductions at lower than expected cost. This success story undoubtedly will encourage Congress and the EPA to use economic instruments more widely in the future. Indeed, many are now calling on Congress to radically transform environmental policy, advocating abandonment of command-and-control in favor of the next generation of efficiency-enhancing market-based controls (see, e.g., Enterprise for the Environment, 1998; Chertow & Esty, 1997).

Rena Steinzor has cautioned against such a radical transformation: "without dramatically expanding the resources available to federal and state regulators, and without placing challenging, new demands on pollution sources to track emissions and research their toxicological effects, the shift to the 'next generation' of regulatory policy is likely to result in severe degradation of environmental quality" (Steinzor, 1998, pp. 10361, 10362).[21] Our analysis suggests that caution is, indeed, warranted. As we have seen, command-and-control mechanisms have reduced air pollution, and they have done so (for the most part) efficiently. Moreover, market-based solutions are not well-suited for all institutional and technological contexts, particularly where monitoring costs are exorbitant. In that circumstance, command-and-control regulations may be both more effective and more efficient. To the extent they are replaced by market mechanisms, it should only be after careful, case-by-case examinations of expected costs and benefits, including implementation and monitoring costs.

On the other hand, our historical analysis of the Clean Air Act also suggests that the radical policy changes Steinzor fears are unlikely to occur. Like other institutions in society, environmental protection (including the regulatory regime itself) evolved slowly, incrementally, and inconsistently. And in this case there is no reason to anticipate deviation from that tendency because the Clean Air Act, as it is, continues to provide substantial (and, apparently, increasing) net benefits for society. Consequently, there is little substantive (as opposed to political or ideological) reason for Congress to radically amend the Act.[22]

Not all environmental laws have been as efficient as the CAA. By all accounts, the costs of the Clean Water Act (another predominantly command-and-control regime), for example, have outweighed its benefits.[23] But it is one thing to note that command-and-control-based regulatory regimes are sometimes inefficient; it is another to assert that they are inherently so.[24] We do not dispute the former assertion, but our analysis suggests that the latter assertion is erroneous.

So, the question becomes, how and when should policy makers employ command-and-control rather than market mechanisms? As Stavins notes, "[t]here is no simple answer, no policy panacea. Inevitably, case-by-case examinations are required" (Stavins, 1993, p. 29). In large measure, the choice of regulatory regime depends on the goals and concerns of policy makers (see generally Office of Technology Assessment, 1995). However, our analysis suggests that where abatement costs are relatively low and monitoring costs are relatively high, command-and-control is likely to be at least as efficient (and effective) as effluent taxes or a tradable emissions program. In the obverse case of relatively high abatement costs and relatively low monitoring costs, market mechanisms are likely to be more efficient. But, of course, there are many other economic, institutional, and technological variables that can affect the comparison of regulatory options, which is precisely why case-by-case examinations are required.

Notes

1. It is also sometimes argued that market-based approaches are more democratic than command-and-control regulations (see, e.g., Ackerman & Stewart, 1985, p. 1334; Stewart, 1992, p. 547). But empirical studies suggest the democratic case for market-based environmental regulation is weak (see Heinzerling, 1995).

2. Throughout this chapter, the labels economic instruments and market-based approaches to regulation are used interchangeably, in contrast to command-and-control. For a primer on the various instruments that fall into these categories, see Stavins & Whitehead (1996).

3. II Legislative History (1974, p. 1306–1307). For descriptions of the technologies available circa 1970 for monitoring air pollution, see Lieberman & Schipma (1969).

4. NAPCA reported to Congress that some industries simply refused to provide requested information about emissions (II Legislative History, 1974, p. 1312).

5. See CEQ (1971, pp. 27, 30). It is interesting to note both SO_2 emissions (in 1990) and lead additives in gasoline (in 1982) were later subject to tradable permitting.

6. A criteria pollutant is one for which the EPA has established national ambient air quality standards under the Clean Air Act. Today, there are six such criteria air pollutants: particulate matter, sulfur dioxide, carbon monoxide, nitrogen oxides, and lead.

7. The legislative history included studies that suggested large cost advantages of effluent taxes over command-and-control regulations. See (III Legislative History, 1978, pp. 1192–1205).

8. Freeman's assessment was subsequently confirmed by Portney (1990, pp. 27, 69).

9. On the experimental nature of emissions trading and other new approaches to environmental protection, see Selmi, 1994, p. 1061.

10. Each allowance equals one ton of SO_2.

11. It is worth noting that these benefit estimates of the acid rain program do not include difficult to quantify environmental benefits, such as reduced acid rain damage to forests, lakes, rivers, and buildings.

12. Calculated from figures presented in Council on Environmental Quality (1990, pp. 320–322, Table 39).

13. Calculated from figures presented in Council on Environmental Quality (1990, p. 323, Table 40). Five of the six criteria pollutants experienced drops in ambient concentration levels. The exception was ozone, ambient concentrations of which increased by seven percent. See Council on Environmental Quality (1990, pp. 320–322, Table 39).

14. On emissions rates in 1977 and 1990, see Council on Environmental Quality (1990, pp. 320–323, Tables 39 & 40).

15. Although precise cost and benefit figures are debatable, according to Paul Portney, "any analysis of the act would conclude that its benefits outweigh its costs." Quoted in Zacaroli (1998). After conducting an extensive review of the literature, Clarence J. Davies and Jan Mazurek concur: "Taken as a whole, the benefits of the Clean Air Act seem clearly to outweigh the costs" (Davies & Mazurek, 1998, p. 278).

16. Indeed, the lack of information about the costs and benefits of the Clean Air Act led Congress, in the 1990 Amendments, to require EPA to report to it on costs and benefits. EPA published its final report in 1997. See U.S. Environmental Protection Agency (1997).

17. EPA's rules were first subject to "Quality of Life" reviews by President Nixon's Office of Management and Budget (U.S. Environmental Protection Agency, 1987, S-2). Subsequently, the agency was required to perform economic analyses of its major rules under executive orders issued by Presidents Carter, Reagan, and Clinton. See Davies & Mazurek 1998, p. 32).

18. For instance, in 1998 the EPA was still working out the bugs in a CEMS for particulate matter emissions from hazardous waste incinerators. See "Hazardous Waste" (1998).

19. We have not attempted in this paper to explain why air pollution control policy in general has been growing more rather than less efficient, despite its continued heavy reliance on command-and-control. As far as we are

aware, no other study has even pointed out that this is the case, let alone explained it. Among the possible reasons are population growth, increasing health care costs, and improvements in methodologies for calculating environmental costs and, particularly, benefits.

20. Of course, high monitoring costs do not always render market-based approaches, such as effluent taxes, less efficient than command-and-control regulations. See Harrington, McConnell, & Alberini (1996).

21. Also, see Davies & Mazurek (1998), arguing that the very distinction between command-and-control and incentive-based approaches to regulation is misguided.

22. It is worth noting that when the 104th Congress attempted a radical revolution in environmental regulation, it was left licking its wounds (Burkley, 1996, pp. 677, 678).

23. See Freeman (1990, pp. 97, 125–126), asserting that in 1985 the annual estimated costs of the CWA ranged between $25 and $30 billion while its most likely estimate of benefits were between $5.7 and $27.7 billion. It is worth wondering, however, whether the costs of water pollution control overall—the Clean Water Act plus the Safe Drinking Water Act—exceed the benefits.

24. Moreover, it is presumptuous to assume efficiency is the lone or predominant goal of environmental legislation (whether or not it should be).

References

Ackerman, B. A., & Stewart, R. B. (1985). Reforming Environmental Law. *Stanford Law Review* 37:1267–1332.

Air Pollution: Acid rain allowance auction generates $32 million in sales; prices up dramatically. (March 28, 1997). *BNA Chemical Regulation Daily.* Available in LEXIS, News library, Curnws file.

Burkley, W. R. (1996). Environmental reform in an era of political discontent: Introduction. *Vanderbilt Law Review* 49:677–688.

Burtraw, D., Krupnick, A., Mansur, E., Austin, D., & Farrell, D. (1997). The costs and benefits of reducing acid rain. Resources for the future discussion paper, No. 97-31-REV.

Council on Environmental Quality. (1990). Environmental quality: The twentieth annual report of the Council on Environmental Quality together with the president's message to congress. Washington, DC: Government Printing Office.

Council on Environmental Quality. (1971). The president's environmental program. Washington, DC: Government Printing Office.

Dales, J. H. (1968). *Pollution, property, and prices.* Toronto: University of Toronto Press.

Davies, C. J., & Mazurek, J. (1998). *Pollution control in the United States: Evaluating the system.* Washington, DC: Resources for the Future.

Energy Information Administration. (1997). *The effects of Title IV of the Clean Air Act Amendments of 1990 on electric utilities: An update.* Washington, DC: U.S. Department of Energy.

Enterprise for the Environment. (1998). *The Environmental Protection System in transition: Toward a more desirable future.* Washington, DC: Center for Strategic and International Studies.

Environmental Protection Agency. (1997). *National air pollutant emission trends, 1900–1996.* Washington, DC: Government Printing Office.

Chertow, M. R., & Esty, D. C. (Eds.). (1997). *Thinking ecologically: The next generation of environmental policy.* New Haven: Yale University Press.

Freeman, A. M. (1982). *Air and water pollution control: A cost-benefit assessment.* New York: Wiley.

Freeman, A. M. (1990). Water pollution policy. In P. R. Portney (Ed.). *Public policies for environmental protection.* Washington, DC: Resources for the Future, pp. 97–150.

Gerhardt, P. H. (1969). Incentives for air pollution control. In C. C. Havighurst (Ed.). *Air Pollution Control.* Dobbs Ferry, NY: Oceana Pubs, pp. 162–172.

Hagevik, G. (1969). Legislating for air quality management: Reducing theory to practice. In C. C. Havighurst (Ed.). *Air Pollution Control.* Dobbs Ferry, NY: Oceana Pubs., pp. 173–202.

Hahn, R. W., & Hester, G. L. (1989). Marketable permits: Lessons for theory and practice. *Ecology Law Quarterly* 16:361–406.

Harrington, W., McConnell, V., & Alberini, A. (1996). Economic incentive policies under uncertainty: The case of vehicle emission fees. Resources for the Future Discussion Paper No. 96-32.

Havighurst, C. C. (Ed.). (1969). *Air Pollution Control.* Dobbs Ferry, NY: Oceana Pubs.

Hazardous waste: New PM continuous emission data show problems with monitoring units. (June 29, 1998). *BNA National Environment Daily.* Available in LEXIS, News library, Curnws file.

Heinzerling, L. (1995). Selling pollution, forcing democracy. *Stanford Environmental Law Journal* 14:300–344.

A Legislative History of the Clean Air Act Amendments of 1990. (1993) Washington, DC: Government Printing Office.

I A Legislative History of the Clean Air Act Amendments of 1970. (1974). Washington, DC: Government Printing Office.

II A Legislative History of the Clean Air Act Amendments of 1970. (1974). Washington, DC: Government Printing Office.

III A Legislative History of the Clean Air Act Amendments of 1977: A Continuation of the Clean Air Act Amendments of 1970. (1978). Washington, DC: Government Printing Office.

IV A Legislative History of the Clean Air Act Amendments of 1977: A Continuation of the Clean Air Act of 1970. (1978). Washington, DC: Government Printing Office.

VI A Legislative History of the Clean Air Act Amendments of 1977: A Continuation of the Clean Air Act Amendments of 1970. (1978). Washington, DC: Government Printing Office.

Janke, J. A. (1993). *Continuous Emission Monitoring.* New York: John Wiley & Sons.

Keohane, N. O., Revesz, R. L., & Stavins, R. N. (1997). The positive political economy of instrument choice in environmental policy. Resources for the Future Discussion Paper No. 97-25.

Krier, J. A. (1974). The irrational national air quality standards: Macro- and micro-mistakes, *UCLA Law Review* 22:323–342.

Lieberman, A., & Schipma, P. (1969). *Air pollution-monitoring instrumentation: A survey.* Washington, DC: Technology Utilization Division, National Aeronautics and Space Administration.

Liroff, R. (1986). *Reforming air pollution regulation: The toil and trouble of EPA's bubble.* Washington, DC: Conservation Foundation.

Magat, W. A. (Ed.). (1982). *Reform of environmental regulation.* Cambridge, MA: Ballinger Publishing Co.

National Academy of Sciences. (1977). National Research Council, *Environmental Monitoring, Vol. IV of Analytical Studies for the US Environmental Protection Agency.* Washington, DC: National Academies Press.

National Commission on Air Quality. (1981). *To Breath Clean Air.* Washington, DC: Government Printing Office.

North, D. C. (1990). *Institutions, institutional change, and economic performance.* Cambridge: Cambridge University Press.

Office of Technology Assessment. (1995). *Environmental policy tools: A user's guide.* Washington, DC: Office of Technology Assessment, Congress of the United States.

Orts, E. (1995). Reflexive environmental law. *Northwestern University Law Review* 89:1227–1340.

Percival, R. V., Miller, A. S., Schroeder, C. H., & Leape, J. P. (1969). *Environmental regulation: Law, science, and policy.* Gaithersberg, NY: Aspen Law & Business.

Portney, P. R. (1990). Air pollution policy. In P. R. Portney (Ed.), *Public policies for environmental protection.* Washington, DC: Resources for the Future, pp. 27–96.

Portney, P. R. (Ed.). (1990). *Public policies for environmental protection.* Washington, DC: Resources for the Future.

Roberts, M. (1982). Some problems of implementing marketable permit schemes: The case of the Clean Air Act. In Magat, W. A. (Ed.). *Reform of environmental regulation.* Cambridge, MA: Ballinger Publishing Co., pp. 93–117.

Russell, C., Harrington, W., & Vaughn, W. J. (1986). *Enforcing pollution control laws.* Washington, DC: Resources for the Future.

Selmi, D. P. (1994). Experimentation and the 'New' Environmental Law. *Loyola (L.A.) Law Review* 27:1061–1076.

Stavins, R., & Whitehead, B. W (1996). The next generation of market-based environmental policies. Resources for the Future Discussion Paper No. 97-10.

Stavins, R. (1993). Transaction costs and the performance of markets for pollution control. Resources for the Future Discussion Paper No. QE93-16.

Steinzor, R. (1998). Reinventing environmental regulation: Back to the past by way of the future. *ELR News and Analysis* 28:10361–10372.

Stewart, R. B. (1992). Models for environmental regulation: Central planning versus market-based regulation. *Boston College Environmental Affairs Law Review* 19:547–560.

Stewart, R. B. (1996). The future of environmental regulation: United States environmental regulation: A failing paradigm. *Journal of Law and Commerce* 15:585–596.

Swift, B. (1997). The acid rain test. *Environmental Forum* 14:17–25.

Tietenberg, T. H. (1985). *Emissions trading: An exercise in reforming pollution policy.* Washington, DC: Resources for the Future.

U.S. Environmental Protection Agency. (1997). *The benefits and costs of the Clean Air Act, 1970–1990.* Washington, DC: Environmental Protection Agency.

U.S. Environmental Protection Agency. (1987). *EPA's use of cost-benefit analysis: 1981–1986.* Washington, DC: Environmental Protection Agency.

U.S. General Accounting Office. (1994). *Air pollution: Allowance trading offers an opportunity to reduce emissions at least cost.* Washington, DC: U.S. General Accounting Office.

Wolozin, H. (1969). The economics of air pollution: Central problems. In Havighurst, C. C. (Ed.). *Air Pollution Control.* Dobbs Ferry, NY: Oceana Pubs., pp. 31–42.

Zacaroli, A. (December 16, 1998). Air pollution: Economist blasts EPA cost/benefit report for Clean Air Act; Others support agency. *BNA National Environment Daily.* Available in LEXIS, News library, Curnws file.

Zacaroli, A. (August 12, 1996). Leading the news: Air pollution: Utilities achieve 100 percent compliance with EPA acid rain program, report says. *BNA Daily Environment Report.* Available in LEXIS, News library, Curnws file.

CHAPTER ELEVEN

Conclusion

MICHAEL T. HATCH

As chronicled in the preceding chapters, the range of policy instruments available for environmental protection has broadened significantly in recent decades. In the face of growing disaffection with traditional command-and-control regulation, policymakers have turned to alternative instruments that hold out the promise of better environmental performance without the liabilities that have accompanied the direct regulatory approach. Whether this promise, in fact, has been fulfilled provides the overarching question that informed the analyses found in this volume.

Assessing the usefulness of policy instruments

There are contending views about the relative virtues of individual policy tools. As detailed in chapter 1, the ongoing debates about the usefulness of various instruments are frequently grounded in preferences that elevate one set of evaluative criteria over others. The contributors to this volume have sought to come to some type of judgment about the ability of a policy instrument to improve environmental performance. Given no single criterion is likely to fully capture the usefulness of a policy instrument, however, such judgments have been measured in different currencies or dimensions of performance (environmental effectiveness, economic efficiency, political efficiency, administrative efficacy, and technological innovation). Let us briefly review what the individual case studies have concluded.

It is asserted that eco-labels offer consumers a choice based on ecological considerations, thereby creating incentives for the innovation of more environmentally sound products or production practices. Proponents argue eco-labels are relatively light in terms of the amount of public expenditure, management, and oversight and, as such, are rather easy to introduce and implement. Critics of such voluntary instruments, however, are concerned the increased flexibility allowed by such approaches will lower overall environmental protection; advocates of economic instruments are skeptical that voluntary arrangements can produce economically efficient results.

With over twenty years of operational experience to draw upon, Müller concludes Germany's Blue Angel eco-labeling program has proven to be an ecologically effective instrument through its promotion of environmentally improved products. The degree of ecological effectiveness, however, depends very much on the nature of a product category: products that were part of the day-to-day consumption and purchasing of private consumers proved the most difficult to effectively incorporate into the Blue Angel; inclusion of environmentally friendly products in public tenders was a major driver of the program. Measured in the currency of economic efficiency, Müller's analysis found little difference in the efficiency of eco-labeling relative to other regulatory tools. In terms of political efficiency, however, the Blue Angel program created an avenue for environmental action at a time when German industry was able to successfully oppose legislative command and control efforts. Further, it encouraged innovation and diffusion in a number of products and technologies and empowered forerunner companies having difficulties placing their products in markets dominated by strong and established economic actors. Finally, Müller persuasively counters the common perception that eco-labeling is a relatively light policy instrument. Far from being a simple self-running and self-financing tool requiring few public resources, the Blue Angel entails a set of complicated procedures and decisions that rely on extensive administrative capacity and technical expertise (more on this later). Also running counter to the view of eco-labels as an easy instrument to introduce and implement, the Blue Angel had to overcome substantial resistance, not only from industrial associations but the consumer community and environmental groups as well.

In contrast to this generally positive assessment of eco-labeling, Lipschutz's overall conclusions on the Forest Stewardship Council's (FSC) efforts to introduce independent eco-labeling of forest products are much more circumspect. Since the long-term consequences of certification can not yet be accurately gauged, the ultimate question about efficacy of private self-regulation through such instruments as eco-labeling was left open. He did, however, find indications the FSC system did not lead to the desired ecological outcome, sustainable forest management. Most important in this regard were questions about whether or not FSC standards are actually being implemented. Suggested among the factors contributing to this were the weak institutional bases of the FSC (scarce funding and personnel to monitor implementation), few penalties for failing to observe the rules, and the introduction of competing programs by forest products companies that adopt a less rigorous approach to standard setting. Similar conclusions were drawn when applying the other criteria. In other words, efforts to privatize regulatory responsibilities in the international arena have met with considerably less success than national eco-labeling programs that enjoy the institutional backing of national governments.

Like eco-labeling, eco-audits are voluntary arrangements that provide consumers with information about environmental management practices. For firms that choose to adopt specified standards for environmental management, certified participation in such programs is designed to foster better relations with customers, suppliers, stakeholders, and employees. Moreover, through this process of self-evaluation, firms discover ways of doing things more efficiently, thereby reducing their ecological footprint. From the perspective of the public sector, such arrangements promise environmental benefits without the high administrative costs that accompany direct regulation. There are, however, concerns that this type of instrument will not actually result in improved environmental protection or that it can produce economically efficient results.

In their analysis of ISO 14001 and the widespread adoption of this environmental management system by Japanese corporations, Welsh and Schreurs found that ISO 14001 certification resulted in significant process and structural change that may be significant for long-term environmental performance. They also found ISO firms were more environmentally active than non-ISO firms. What they were not able to do, however, was establish a direct causal relationship between ISO and improved environmental behavior. What seemed to explain the high levels and rates of adoption of ISO 14001 in Japan were the importance Japanese firms attached to presenting a green image to the international community and domestic consumers along with an institutional setting that facilitated the adoption of ISO 14001 and kept the costs of administering the program relatively low.

Voluntary agreements, it is argued, provide firms much greater flexibility in terms of method and timing, thus helping to achieve environmental objectives at lower costs as well as stimulating innovation. Given the more collaborative nature of the policy process, such an approach is said to promote faster implementation and compliance. Critics, on the other hand, argue they are unlikely to move beyond what industry would have done in their absence and, moreover, they will fall well short of what could have been achieved with the adoption of formal regulations.

According to the analysis of voluntary agreements between the German government and the power generation sector to limit CO_2 emissions, the commitments evolved from rather weak to quite ambitious quantified targets. In other words, they held the promise of improved environmental effectiveness. But since those target dates are still several years away, a more definitive judgment on the environmental effectiveness of the instrument awaits the translation of commitments into concrete achievements. Nonetheless, a key to enhancing the potential effectiveness of the agreements was, according to Hatch, the adoption of monitoring/reporting procedures along with an independent assessment system. Also significant was the political efficiency of the voluntary agreements in that they provided an alternative to the political impasse over climate change policy in Germany during the mid-1990s. Technology-forcing aspects of the voluntary agreements, on the other hand, were ambiguous: given the modesty of initial commitments, they were unlikely to affect technological choices, but the latter agreements were designed to encourage the expansion of cogeneration and fuel cells.

Proponents of green taxes argue that firms—when faced with the direct costs of their polluting activities—have incentives to control pollution. At the same time, they are free to choose the most efficient reduction methods. Green taxes also are said to provide ongoing incentives to find the most efficient reduction technologies in order to lower or avoid the tax. Finding the proper level of taxation, however, is critical to the effectiveness of the instrument. According to critics, however, this is especially problematic since—when implemented in the real world—tax levels reflect the influence of powerful political forces as much as market forces. This limits the environmental effectiveness—as well as economic efficiency—of the instrument.

Though Germany's ecological tax reform (ETR) has only been in effect since 1999, Kohlhaas and Meyer nonetheless found the ETR has been successful in reducing CO_2 emissions. It has not been optimally effective in achieving environmental ends, however, due to legal and political constraints that lead to inconsistencies in patterns of taxation (for example, exemptions for energy-intensive sectors; withholding of exemptions for power generated from renewables). Those same inconsistencies impair the economic efficiency of the ETR as well, but Kohlhaas and Meyer concluded it remains a cost-efficient instrument that would not have been adopted in the absence of a political process able to balance competing objectives, interests and legal exigencies. In other words, the ETR is the result of a politically efficient process able to bring together different constituencies representing separate environmental and economic concerns. Finally, in response to the introduction of the ETR, many producers offer more energy-saving products (e.g. car manufacturers), indicating the potential of the ETR to spur technological innovation.

By using the price mechanism to internalize the costs of pollution, emissions trading is said to encourage efficiencies that lead to ongoing pollution reduction. Specifically, they generate efficiencies by reducing compliance costs and providing greater flexibility. On the other hand, critics have voiced concerns about the normative issues surrounding the sale of the right to pollute and environmental justice.

Through an analysis of several emissions trading arrangements in the United States, Bryner found that, given the proper framework conditions, emissions trading programs represent a cost-effective means to accomplish goals. Specifically, they generate efficiencies by reducing compliance costs and by providing greater flexibility. He argues the goals themselves may not be sufficient to remedy the problems and emissions trading programs may lead to high concentrations of pollutants in some areas, raising questions about environmental effectiveness.

Bryner demonstrates that emissions trading helps generate political support for ambitious new regulatory programs and opens up possibilities for new political coalitions; and in terms of administrative efficacy, Bryner argues emissions trading simplifies some of the regulatory tasks of government agencies and compliance burdens of regulated sources. Finally, Bryner shows emissions trading encourages the innovation and diffusion of control technologies, though he cautions that unless trading programs are aggressively structured, they may reduce pressures to innovate and result in the purchase of credits rather than investment in pollution control equipment.

Oberheitmann's discussion of the JI and CDM emissions trading arrangements proposed in the Kyoto Protocol make clear the importance of providing more cost-effective options if international efforts to address the challenge of climate change are to receive broad acceptance. It also highlights the reluctance of private companies to participate in these flexible instruments and certain measures national governments might find necessary to encourage private sector participation.

Environmental impact statements (EISs) represent a process designed to identify, evaluate, predict, and mitigate the effects of proposed developments before decisions are taken or projects initiated. Generally mandated by law, they provide information on environmental effects, risks, and consequences of development proposals. By explicitly integrating science into the decisionmaking process, EISs facilitate the inclusion of environmentally sound and sustainable options in proposed projects. On the other hand, critics claim the EIS process is costly and time-consuming.

According to Rosenbaum, EISs were adopted in the United States with the intention to create a profound change in federal environmental policymaking. While perhaps not meeting the high expectations of their proponents, Rosenbaum argues that IESs have a significant impact on environmental policymaking in the United States. Most importantly, the IES process is a powerful stimulant to the growth of ecological science and various scientific disciplines, thereby enhancing the quality and use of ecological information. Whether or not this results in improved environmental conditions is difficult to determine, but Rosenbaum found sufficient evidence to suggest federal agencies frequently made different, and often ecologically better, decisions as a result of the EIS process. In terms of economic efficiency, there is little evidence to suggest EIS procedures imposed excessive costs or delay upon agency decision making or that a significant number of federal projects were casualties of NEPA-imposed reviews, though estimates of the benefit/cost value of the EIS process itself are almost never calculated.

The political effect of the EIS process has been to enlarge the number of interests active in agency decision making. At the same time, Rosenbaum finds EISs are believed to encourage greater acceptance of related agencies' policies and facilitate the subsequent implementation of those policies. Finally, in conjunction with technological innovation, he argues EIS preparation is a major reason for the development of the relatively new science of ecosystem analysis and for increased research in the cumulative effects of environmental interventions.

Proponents of command and control instruments argue they are effective in achieving their specified environmental objectives. An added benefit of technology-based regulations is that compliance and oversight are much easier. The most common critiques of traditional command and control regulatory instruments focus on their inability to effectively address certain types of environmental problems along with their high cost and inefficiency.

Taking to task proponents of economic instruments (who tend to ignore the type of framework conditions provided in Bryner's analysis), Cole and Grossman argue command-and-control instruments can be and have been more efficient and effective than economic forms of regulation. Critical in this regard are technological and institutional factors: where abatement costs are relatively low and monitoring costs relatively high, command and control instruments are at least as efficient and effective as effluent taxes or a tradable emissions program. On the other hand, where relatively high abatement costs and relatively low monitoring costs prevail, market mechanisms are likely to be more efficient. In conjunction with this argument, Cole and Grossman demonstrate that technology-based design standards facilitate implementation since they require extensive monitoring and staffing capabilities; technological changes (e.g., continuous emission monitoring systems) subsequently made tradable permits and effluent taxes administratively feasible.

In sum, conclusions about the efficacy of the policy instruments included in this volume are somewhat mixed. Of necessity, several are more tentative than others, due to either the relatively short period of time the policy instruments have been in place and/or limitations on the data required to establish causal links. Nonetheless, the findings are largely positive:

- Though judgments were at times hedged or qualified (e.g., for eco-labeling, the nature of the product category; for emissions trading, the aggressiveness of the goals) each of the instruments was found to be either environmentally effective or potentially so, the major exception being forestry eco-labeling programs.
- Concerns about economic efficiency are central to arguments favoring market-based instruments. Both emissions trading and eco-taxes are found to be cost efficient though, at least in the case of the latter, not optimally so given certain legal and political constraints. As argued by Cole and Grossman in their analysis of the Clean Air Act, however, claims to economic efficiency are not the sole preserve of such market-based instruments.
- As argued in the studies of eco-labels, voluntary agreements, eco-taxes, emissions trading, and EIS, it is important to take into account questions of feasibility, second-best solutions and the possibility of reconfiguring political support. In other words, considerations of political efficiency can and perhaps ought to be included in any realistic assessment of policy instruments.
- Considerations of administrative efficacy were included in the evaluation of several cases, most importantly within the context of facilitating implementation (e.g., the design standards contained in command and control regulations) and compliance (monitoring/reporting arrangements for voluntary agreements; emissions trading).
- Finally, a number of studies demonstrate the potential for technological innovation and diffusion (e.g., eco-labels, eco-taxes, and emissions trading). For others, the technology-forcing aspects of the instrument are more ambiguous (voluntary agreements) or

even malign (e.g., if emissions trading result in the purchase of credits rather than investment in pollution control equipment).

What can we conclude about the ability of various policy instruments to improve environmental performance? There are no simple answers to this question. While the contributors to this volume come to a generally positive assessment of the various policy instruments, it is clear no single criterion fully captures the efficacy of individual policy tools.

Case studies and broader generalizations

As detailed in the foregoing chapters, the choice of instruments is shaped in large part by a set of political, economic, societal, and technological conditions that often differ across nations. Through the use of a case studies approach that focuses on individual policy instruments, we have added our voices to ongoing efforts that seek to identify factors that contribute to the success of those instruments within specific framework conditions. What we are less able to do using this type of approach is generalize about those factors across national settings. Nonetheless, there are several common themes that emerge from the analyses contained in this volume.

Among these themes is one that resonates with observations from other studies in environmental policymaking (see, for example, Keohane, Haas, & Levy, 1993, referenced in chap. 1), the importance of institutional capacity for the effective formulation and implementation of environmental policy:

- In her examination of the Blue Angel program, Müller persuasively counters the common perception that eco-labeling is a relatively light policy instrument in that the amount of public expenditure, management, and oversight are minimal. Far from being a simple tool, implementation of the Blue Angel entails a set of complicated procedures and decisions that rely on the extensive administrative capacity and technical expertise provided by public authorities along with private experts and stakeholders in the process of selecting product categories, defining and elaborating the awarding criteria, examining the applications of manufacturers and certifying their products.
- In contrast, the ineffectiveness of the Forest Stewardship Council's certification standards, according to Lipschutz, is due to its weak institutional bases, in particular scarce funding and personnel to monitor compliance.

- As detailed in the Bryner and Cole/Grossman studies on efforts to address acid rain pollution in the United States, monitoring and staffing are cited as critical factors in the choice—as well as the effectiveness—of policy instruments. Given the lack of air pollution information due to existing technological constraints, a command-and-control regulatory approach was initially employed. Once the quality and reliability of monitoring equipment, analytical techniques, and capabilities were improved, however, the use of emissions trading became more feasible.

- In the case of voluntary agreements to reduce CO_2 emissions in Germany, Hatch concludes the decision to establish monitoring and evaluation capabilities was critical in providing the impetus toward a more environmentally effective instrument.

A related theme common to several case studies is that policy instruments seldom function as effectively when standing alone. Müller, for example, argues that the impact of eco-labeling in Germany was improved substantially when supported by and complemented with additional tools and measures (e.g., a green public procurement policy by government). In trying to explain why the adoption of ISO 14001 by Japanese firms is so pervasive, Welsh and Schreurs emphasize the importance of the support provided by environmental laws and elements of Japan's public sector.

A third point is that certain policy instruments act as precursors to or facilitators for the introduction of other instruments:

- For Bryner, emissions trading serves as a transition to a more effective set of market-based regulatory instruments. In his view, the real importance of trading programs lies in their ability to generate support for the idea of market-based incentives, provide data on ecological costs and benefits, and pave the way for more fundamental policy reforms.

- Müller argues that when such soft tools as eco-labels are found inadequate, this can provide arguments and scope for actions to introduce other measures.

- Similarly, part of the debate surrounding the adequacy of voluntary agreements in meeting Germany's CO_2 reduction targets is the prospect of more direct regulation (should the results prove disappointing). Moreover, much of the detailed information necessary for this type of regulatory approach has been generated by the reporting requirements and evaluation process industry submits to under the voluntary agreement.

One final point touches on the sample of case studies found in this volume. A possible critique is that the analyses focus primarily on the experiences of large, post-industrial democracies. While true, there are benefits to such an approach, given the influence these case studies have had on the diffusion of these policy instruments:

- From its inception in the United States, the use of environmental impact assessments has spread to well over one hundred countries.
- The U.S. experience in emissions trading has provided valuable guidance in the creation of emissions trading programs throughout the world.
- In recent years, Japan has become a global leader in the number of ISO certified companies, a practice that has been transferred throughout the world as Japanese firms encourage their foreign affiliates to seek ISO 14001 certification.
- For a decade, the Blue Angel was the only eco-labeling program in the world; today, there are approximately thirty countries using environmental labels, virtually all have been influenced to some degree by the German program.

While the role of small countries as innovators or forerunners in environmental policy should not be minimized, one might ask whether the diffusion of these policy instruments would be so extensive if these countries were not among the most politically and/or economically influential in the world.

To conclude, these case studies contribute valuable insights into the functioning of policy instruments within specific institutional settings and, as such, provide important reference points for further study. More broadly, the common themes emerging from these studies suggest that policy instruments, standing alone, rarely represent a silver bullet in the arsenal available to policymakers in the environmental arena. They are most effective when employed in conjunction with other policies or measures. Given the relatively limited experience with many of these instruments to date, however, such findings caution against premature efforts to harmonize or seek uniformity in the policy instruments employed. Rather, experimentation with various instruments used in combination with other policies or measures provide the kind of approach needed in efforts to manage the complex interactions of humans with their physical environment.

Contributors

Gary C. Bryner is Professor of Political Science at Brigham Young University, where he has directed the Public Policy Program. He was Director of the Natural Resources Law Center and Research Professor of Law at the University of Colorado School of Law. He is the author of *Blue Skies, Green Politics: The Clean Air Act of 1990 and Its Implementation; From Promises to Performance: Achieving Global Environmental Goals;* and, with Jacqueline Vaughn Switzer, *Environmental Politics: Domestic and Global Dimensions.*

Daniel H. Cole is the M. Dale Palmer Professor of Law at the Indiana University School of Law at Indianapolis. Professor Cole teaches and writes about the law and economics of Property and Environmental Protection. He writes extensively about Poland and Polish law. Professor Cole is the author or editor of two books and twenty-five articles, book chapters, and essays. His recent book, *Instituting Environmental Protection: From Red to Green in Poland* received the 1999 AAASS/Orbis Polish Book Prize.

Peter Z. Grossman is the Clarence Efroymson Chair in Economics, College of Business Administration, Butler University. He has specialized in the fields of law and economics and economic history. Prior to assuming the Efroymson Chair, Dr. Grossman taught at Polytechnic University (New York) and at Washington University (St. Louis). He has also been affiliated with the Center for Political Economy and the Center for the Study of American Business, both at Washington University. Dr. Grossman's work has looked primarily at issues of industrial organization, energy policy and environmental regulation. He has published three books and many articles, including *Introduction to Energy, Resources, Technology and Society,* a book on energy and public policy coauthored with Edward S. Cassedy. He is currently working on a book (along with Daniel H. Cole) entitled *The End of Natural Monopoly: Deregulation and Competition in the Electric Power Industry.*

Michael T. Hatch is Professor of Political Science at the University of the Pacific, where he teaches courses on European politics, international politics, and global environmental policy. His current research interests include comparative energy and environmental policy, with a focus on the role of domestic politics in international negotiations. Among his publications is *Politics and Nuclear Power: Energy Policy in Western Europe.*

Michael Kohlhaas is a senior economist at the German Institute for Economic Research, specializing in environmental economics. He has undertaken research on ecological tax reform since 1993 and has published four major studies and several papers on the topic. Other fields of interest are voluntary agreements and tradable emission permits as well as economic modeling.

Ronnie D. Lipschutz is Professor of Politics at the University of California, Santa Cruz. His most recent books are *After Authority—War, Peace, and Global Politics in the 21st Century* and *Cold War Fantasies—Film, Fiction, and Foreign Policy.* He has also written a number of books and articles about global environmental politics.

Bettina Meyer is in charge of climate protection and financial aspects of environmental policy in the Ministry of Environment, Nature and Forestry in the state of Schleswig-Holstein. She previously worked as scientific advisor on environmental tax reform for the Green Group in the Federal German Parliament. In cooperation with the Federal Environmental Ministry she is still involved in the process of designing and forwarding the ecological tax reform (ETR) in Germany. Bettina Meyer has written several papers on controversial issues of ETR, for example on tax exemption for renewable energies, and on tax concessions for energy intensive firms.

Edda Müller is the "Mother of the Blue Angel." She introduced the Blue Angel Program in Germany in 1978 and managed it for 16 years. She is the Executive Director of the Federation of German Consumers. She has also served as Deputy Director at the European Environment Agency, Director for Climate Policy at the Wuppertal Institute for Climate, Environment and Energy, Minister of Nature and Environment in the state of Schleswig-Holstein, and Deputy Director General, the German Federal Ministry for Environment, Nature Conservation and Nuclear Safety. Among her publications is *Innenpolitik Der Umweltpolitik (The Inside of Environmental Policy),* 1995 (2nd ed.).

Andreas Oberheitmann is a senior research fellow in the department of energy at the Rhine-Westphalian Institute of Economic Research, Essen, Germany. His research activities and publications cover such areas as the global climate change and the Kyoto process, European Union energy policy, energy and environmental policy in China, and the relationship of China to the World Trade Organization. He is currently working on the Chinese WTO-accession and its impact on the energy sector and the Chinese environmental policy.

Walter A. Rosenbaum is Professor of Political Science at the University of Florida. He also serves as a Consultant for the Office of the Executive Director, South Florida Environmental Restoration Task Force and the U.S. Department of Energy, Office of Environmental Restoration and Waste Management. His publications include *The Cold War's Last Legacy: Risk, Public Opinion and Nuclear Waste Policy,* coedited with James L. Regens. MIT Press (2002) and *Environmental Politics and Policy,* 4th ed. Congressional Quarterly Press, 1998.

Miranda A. Schreurs is Associate Professor in the Department of Government, University of Maryland at College Park. Her current research interests focus on comparative environmental and energy politics in East Asia and the European Union. Among her publications are: *Environmental Politics in Japan, Germany, and the United States; Ecological Security in Northeast Asia* (Seoul: Yonsei University Press, 1998) coedited with Dennis Pirages and *The Internationalization of Environmental Protection* (Cambridge: Cambridge University Press, 1997) coedited with Elizabeth Economy.

Eric Welch is an Assistant Professor in the Graduate Program in Public Administration at the University of Illinois at Chicago. During the last few years, in cooperation with researchers at the National Institute for Environmental Studies in Japan, Dr. Welch has investigated patterns of business greening in Japan, private sector adoption of the ISO 14001 environmental management system in Japan, and the evolution and effectiveness of the Japanese system of pollution control agreements. He has published articles in journals such as *Public Administration Review, Journal of Policy Analysis and Management,* and *Environmental Science and Policy.*

Index